Quantifying Measurement

The tyranny of numbers

Quantifying Measurement

The tyranny of numbers

Jeffrey H Williams

Morgan & Claypool Publishers

Rights & Permissions
To obtain permission to re-use copyrighted material from Morgan & Claypool Publishers, please contact info@morganclaypool.com.

ISBN   978-1-6817-4433-9 (ebook)
ISBN   978-1-6817-4432-2 (print)
ISBN   978-1-6817-4435-3 (mobi)

DOI   10.1088/978-1-6817-4433-9

Version: 20161001

IOP Concise Physics
ISSN 2053-2571 (online)
ISSN 2054-7307 (print)

A Morgan & Claypool publication as part of IOP Concise Physics
Published by Morgan & Claypool Publishers, 40 Oak Drive, San Rafael, CA, 94903 USA

IOP Publishing, Temple Circus, Temple Way, Bristol BS1 6HG, UK

For BTC, without whom none of this would have been possible.

Contents

Introduction

Just about everyone who has ever studied science has done an experiment. Theories have their place, but they are like fashions in that they change with time, and are only of relevance until such time as someone devises an experiment to test their veracity. Then, the results of a well-designed experiment last forever; just think of Archimedes in his bath, or Galileo on the leaning Tower of Pisa or Isaac Newton and his prism in a darkened room containing only a shaft of sunlight. Experimentation is the essential motor that drives the advance of science. This dominance of the experimental result as the final arbiter in a consideration of the possible theoretical models of behaviour is as true in particle physics as it is in biochemistry and medicine. Indeed, those branches of science that are most relevant to our society involve only experiments. Our health and civilization are based on the results of experiments, and we train future scientists by getting them to undertake established experiments, and then to design their own, novel experiments. But how do you design an experiment, and what should your considerations be on deciding whether an experiment has yielded the sought after, or indeed any useful result?

This might all seem self-evident; however, it is true to say that an appreciation of the importance of the quantification of measurement is not as widespread as it should be. Too many reports of experiments leave too much unexplained or unappreciated, and the errors and levels of precision quoted often suggest that they have been estimated purely as an afterthought. The truth is that there is always an uncertainty in each and every experiment.

Imagine a set of data consisting of measurements of something plotted against time. The data is very 'noisy' with a wide-scatter of points, with the line of data points zigzagging up and down wildly, but there is perhaps a suggestion, or a hint of a general downward trend with time. If this plot represented the number of people out of work, you could be the Prime Minister seeking justification for your government's economic policies; on the other hand, if the data represented the polarization state of the cosmic microwave background radiation, you could be a cosmologist seeking evidence for gravitational waves, yet the problem is the same in both instances. That is, trying to find the needle of useful data you believe to be hidden in the haystack of obscuring, background noise.

All experiments involve trying to measure the thing of interest against a background of unwanted noise, which can have a multitude of sources. Sometimes the amount of noise is small compared to what it is that you wish to measure, and the desired result is clearly seen and the measurement readily made. However, perhaps most frequently, the level of noise is greater than the signal of interest, and various techniques must be adopted to permit the detection system to discriminate between the signal and the noise. This is the basis of experimental design, which allows one to determine the limiting sensitivity of the proposed experiment, and so to answer the questions, is this apparatus worth building to do the experiment that interests me, or is there a different design that leads to a more sensitive apparatus?

Lord Rutherford famously said, 'If your experiment needs statistics, you ought to have done a better experiment.' There is some truth in this critique of experimental design, but an analysis using statistics of your measurements can also be considered as a useful post-mortem examination; the statistical analysis may reveal why the experiment did not work. Often in science, the important thing is not so much the measurement of new facts and figures, or the design of a new experiment, but to discover new ways of thinking about those facts and measurements.

In this volume, we will examine an unexpected by-product of the metric survey of the 1790s. Much is made today, by detractors of the Metric System, of Pierre Méchain's 'error'. This was his response to discovering that his measurements were not as precise as those made by his surveying colleague. Poor Méchain was accused of fudging or manipulating his data (an unpardonable sin as far as scientists were, and are concerned) so as to make it appear that his results were more precise than they actually were. The actual reason for this variation in precision is the subject of this book, and we will see how a detailed analysis of the origins of experimental uncertainly led to the development of the science of statistics, and to the modern method of data analysis. It is not my intention to present a detailed description of the statistical and probabilistic methods used by present day scientists to analyze their measured data, as there are many standard texts and the Internet contains many of these texts as freely accessible pdf files. What I do wish to present in these pages is how we think about experiments, and the difference between a quantitative and a qualitative approach to looking at measurements.

Author biography

Jeffrey H Williams

Jeffrey Huw Williams was born in Swansea, Wales, on 13 April 1956, he gained his PhD in chemical physics from Cambridge University in 1981. His career has been in the physical sciences. First, as a research scientist in the universities of Cambridge, Oxford, Harvard and Illinois, and subsequently as a physicist at the Institute Laue-Langevin, Grenoble, one of the world's leading centres for research involving neutrons and neutron scattering.

Jeffrey Williams has published more than sixty technical papers and invited review articles in the peer-reviewed literature. However, he left research in 1992 and moved to the world of science publishing and the communication of science by becoming the European editor for the physical sciences for the AAAS's Science. Subsequently, he was the Assistant Executive Secretary of the International Union of Pure and Applied Chemistry, the agency responsible for the advancement of chemistry through international collaboration. Most recently, 2003–2008, he was the head of publications at the *Bureau international des poids et mesures* (BIPM), Sèvres. The BIPM is charged by the Metre Convention of 1875 with ensuring world-wide uniformity of measurements and their traceability to the International System of Units (SI). It was during these years at the BIPM that he became interested in, and familiar with the origin of the Metric System, its subsequent evolution into the SI, and the coming transformation into the Quantum-SI.

Since retiring, Williams has devoted himself to writing; in 2014 he published *Defining and Measuring Nature: The make of all things* in the IOP Concise Physics series. This publication outlined the coming changes to the definitions of several of the base units of the SI, and the evolution of the SI into the Quantum-SI. Last year Williams published *Order from Force: A natural history of the vacuum* in the IOP Concise Physics series. This title looks primarily at intermolecular forces, but also explores how ordered structures, whether they are galaxies or crystalline solids, arise via the application of a force.

Chapter 1

The tyranny of numbers

1.1 Why we measure things

Science may be defined in many ways, with as many definitions as there are interested or dis-interested social groups. But those of us with experience of having trained and worked as scientists think of science as the quantitative study of the complex, coupled relationships that may, or may not exist between observed events. Anything that can be measured, anything that can be weighed, anything that can be numbered, anything that can be expressed mathematically—the readings on dials, the 'clicks' and signals coming from a counter or detector can all be considered as part of the enterprise of science. And you can be sure that if you do not measure something, then it will not be included in any scientific analysis, and will likely be lost or forgotten. Today, our scientific knowledge allows us to rationalize the phenomena we see around us, and to make predictions about phenomena as yet unobserved. What was once considered to be magical, is today considered to be rational. Indeed, one could say that magic exists only until it is rationalized by science, when it becomes a banal fact. This is the power of reproducible science; it allows you to be confident about your conclusions.

Conversely, there is no room in the scientific world-view for the inexact, the un-contingent, the immeasurable, the imponderable or the undefined. A process that can be repeated time after time, a system that can be reproduced and analyzed, these are the concepts that go to make up science, and not the individual, the unique, the elusive thing or phenomenon that can never occur a second time. Yet even as some measurements are so complex, or expensive that they cannot be repeated endlessly we must still have confidence in these rare events. Hence our confidence must also rest upon the theory behind the measurement, and on the methodology and practice of the experiment.

But one should also take care not to go to the other extreme and say that if we can measure something, then it must be important; that its significance or utility comes solely from the fact that it is susceptible to measurement. Whether you are part of a

doi:10.1088/978-1-6817-4433-9ch1

1-1

huge team of research physicists observing the Higgs Boson at the Large Hadron Collider at CERN, a technician in a hospital scanning for tumours in thousands of mammographs, or an investor staring at a graph on a computer screen trying to decide whether you should sell your shares today or hang on for another day or so, hoping to make more money, you are all making measurements or observations and are taking decisions, often extremely important decisions, on the basis of those measurements. Each measurement is important to some individual, and the reasons for this personal interest are many and varied.

Measurements are made everywhere, for all of us. This is perhaps best seen in manufacturing and commerce, but the intricate and invisible networks of services, suppliers and communications upon which our society is dependent also rely on metrology for their efficient and reliable operation. International time coordination; for example, involves the most precise of all routine measurements (we can define the second to better than one part in 10^{15}), permitting synchronization of computer networks for: communications, banking, satellite navigation systems that allow accurate location via the Global Positioning System, and among many applications enable aircraft to land in poor visibility and the motorist to find his way home (so precision in time measurement and in the measurement of distance are paramount to our contemporary society). Likewise, our health depends upon accurate diagnosis and the ability to deliver effective treatment based on the precise measurement of quantities of drugs, of ionizing radiation, or of viral nucleic acid in our blood; and the subsequent detailed measurements of the effect of these drugs and radiation on the infectious agents and rogue cell in our bodies.

Many physical and chemical measurements affect the quality of the world in which we live. Incorrect and/or imprecise measurements with regard to the changing environment, or of the levels of pollutants entering the biosphere can lead to the wrong decisions being taken, or to no decision at all being taken by politicians and industrialists, which can have serious consequences, costing a great deal of money and even lives. It is important therefore to have reliable and accurate measurements. It is estimated that in Europe today about six percent of our total gross national products are spent on measurements of one kind or another. Metrology has become, whether we realize it or not, an essential part of our lives. Everything from coffee to concrete, and from recreational drugs to prescribed drugs is bought according to weight, and flows of water, electricity, heat and natural gas are metered. Bathroom scales measure our weight and affect our humour and mood, and police speed traps can affect our finances. The pilot observes his altitude, course and fuel consumption; the authorities monitor the bacterial content of our food and water, and the surgeon primes his laser to cut into our bodies to the nearest fraction of a millimetre. Indeed, it is probably not too much of an exaggeration to say that today it would be difficult to think of anything without referring in some way to weights and measures.

Our world is measured and calibrated, and we are subject to the tyranny and authority of these numbers. This is the way things are, and our education system should train us for living in this quantitative world. To function in society, you need to possess an idea or a sense of how to estimate distance, time, mass and value. If this is not the case, then you will become a victim of; for example, those politicians who are prepared to manipulate statistics to further their own ends at election times.

1.2 A little history

The origin of the system of weights and measures most widely used today can be traced to two events; the creation and implementation of the decimal Metric System in France during the years immediately following the French Revolution, and the development of mass production using interchangeable parts in the industrial revolution. These two events were not, however, directly linked. The Metric System was not created in order to facilitate the mass production of engineered products, and the early development of mass production did not rely upon the new metric or decimal units of measurement. Indeed, the first nations to exploit industrial mass production were the UK and the USA, who used inches and pounds as units in their industries. The Metric System arose from an attempt to unify and bring order to the confusion generated by the multitude of units then in use in France for trade and commerce, and to embrace the grand philosophical concept of constructing a set of units that were in some way derived from nature, and unrelated to material objects or artefacts. The development of mass production, on the other hand, was related to the need to produce as many mechanical devices, or guns, or as much screw-thread as possible in the shortest time.

The Metric System came from the bloodiest period of the French Revolution. Civil war was raging in France while the *savants* were putting the finishing touches to the new Metric System, which became mandatory throughout France in April 1795. Unfortunately for the *savants* who created the Metric System, their splendid philosophical idea was not readily accepted by the ordinary people of France. And it was not until 1840 that the Metric System was finally adopted by France as the sole legal system of measurement. By that time, however, manufacturing industries in the UK and in the USA had already become completely locked into the familiar national standards of the inch and the pound. However, with time the Metric System has become the world's system of weight and measures. The origins of this modern language of science have been presented elsewhere (Williams J H 2014 *Defining and Measuring Nature: The make of all things* (San Rafael, CA: Morgan & Claypool)). Here we will use the origin of the Metric System as the starting point to investigate how scientists quantify measurements; that is, how do we determine what is or is not worth measuring, and how well or how badly has something been measured?

1.3 Surveying

In the 17th Century, *savants* in England and then in France had shown how a new system of weights and measures could be based on a single universal measurement; a measurement of length. This universal length measurement could be defined in terms of the dimensions of the Earth, rather than the length of a monarch's arm or foot, and could then be used to define the other quantities needed by technology; that is, mass, as derived from the weight of a defined volume of pure water.

At the time of the French Revolution, a science commission was set up by the French *Académie des sciences* to determine the practicalities of creating such a new universal system of weights and measures, and this commission recommended a measurement of the new standard of length, the metre, based on a detailed survey along the meridian extending from Dunkirk to Barcelona, which had already been surveyed and measured by the Abbé Nicolas-Louis de Lacaille and César-Francois Cassini in 1739. The commission calculated that if they could measure a significant piece of the meridian, the rest could be estimated. Both ends of the line to be surveyed needed to be at sea level, and as near to the middle of the Pole-to-Equator Quadrant as possible to minimize errors. The meridian chosen is about a tenth of the distance (about one thousand kilometres) from the Pole to the Equator and it runs through Dunkirk, Paris and Barcelona, so most of the distance to be surveyed lay conveniently inside France.

The *Académie des sciences* may have decided that the metre would be exactly a ten millionth of the distance between the North Pole and the Equator, but their choice also defined this distance as being precisely 10 000 000 metres. Unfortunately, an error was made in the commission's initial estimation, because the wrong value was used in correcting for our planet's oblateness. We now know that this Quadrant of the Earth is 10 000 957 metres. One should never forget, that these *savants* were not only setting out to create what they saw as a philosophically coherent system of units based on the dimensions of the Earth, but they were also imposing models and views about the character of the Earth. In 1791, a handful of mathematicians, guided by the writings of Sir Isaac Newton, imposed a definite shape and size to our planet; the Earth shrank and became precisely known.

In the afternoon of 19 June 1791, Jean-Dominique de Cassini (head of the Royal Observatory) had secured an audience with King Louis XVI for some of the members of the metric commission of the *Académie des sciences*. At six in the evening, Cassini, Adrien-Marie Legendre, Pierre François-André Méchain and Jean-Charles de Borda (the inventor of the repeating circle which was hoped would increase the level of precision possible in surveying and thus allow the determination of the metre) presented themselves at the *Palace du Tuileries*. A small group of eminent astronomers and mathematicians had come to convince the now constitutional King Louis XVI that the metre was something worth achieving.

History has not been kind to Louis XVI. He has a reputation for having been naïve and something of a simpleton, but he had hidden talents. The king was a skilled instrument (watch) maker and something of a cartographer. The king also took a close interest in the cost and necessity of the proposed survey. Turning to the head of the Royal Observatory he asked, 'How's that, Monsieur Cassini? Will you again measure the meridian your father and grandfather measured before you? Do you think you can do better than they?' Monsieur Cassini (the third generation of a dynasty of directors of the Royal Observatory) was not unused to conversing with the monarch. 'Sire, I would not flatter myself to think that I could surpass them had I not a distinct advantage. My father and grandfather's instruments could but measure to within fifteen seconds (of a minute of a degree); the instrument of Monsieur Borda here can measure to within one second;' see figure 1.1. (The details of these conversations were published in the *comptes rendue* of the *Académie des sciences*.)

King Louis XVI gave his formal approval for the new survey of the Dunkirk–Barcelona meridian. Then in the early hours of the next day, the king and his family attempted to escape from France (the 'Flight to Varennes'), but they were arrested, returned to Paris and then imprisoned. What must the king have been thinking when

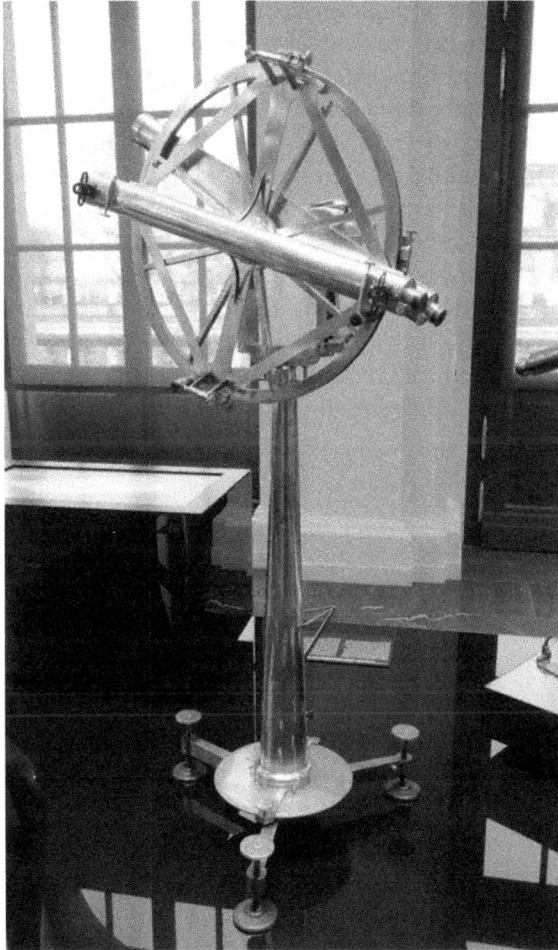

Figure 1.1. A Borda repeating circle; the world's first high-precision measuring device (Exhibit in the Mathematisch-Physikalischer Salon (Zwinger), Dresden, Germany; the image is in the public domain). The first repeating circle was built by Étienne Lenoir in Paris c. 1790; the design was subsequently perfected by Jean-Charles de Borda. Details of the finesse of this instrument may be seen on the Borda circle at the Royal Maritime Museum, Greenwich, at http://collections.rmg.co.uk/collections/objects/42288.html
A surveyor using this instrument would be interested in determining the angle between a cardinal direction and an object: a hill, a church tower, etc. The telescope is trained and centred on the object of interest and the angle between the long axes of the telescope gives the angle that defines the location of the object.

he was contemplating his escape and the suppression of the revolution, while trying to comprehend the mathematics on which he was being lectured by his visitors? However, under arrest or not, Louis XVI was still the king, and from his prison cell he issued the proclamation that directed the two eminent astronomers Jean-Baptiste Delambre (1749–1822) and Pierre Francois André Méchain (1744–1804), to undertake the surveying necessary to determine the length of the metre by precisely measuring the distance from Dunkirk to Barcelona. The king also issued orders to Baron Gaspard Clair François Marie-Riche de Prony to produce new trigonometry tables, with a greater degree of precision, which would be needed to calculate the new universal measure from the surveying work of Delambre and Méchain.

The survey from Dunkirk to Barcelona was a major undertaking. Antoine Lavoisier called it 'the most important mission that any man has ever been charged with'; the measurements were designed to have been completed within a few months, yet it took the two surveyors from May 1792 to September 1798 to complete the work. The technical difficulties were compounded by more practical problems, such as civil and international wars. France was in uproar with some cities restoring governments favourable to the monarchy.

The surveying method used by Delambre and Méchain was triangulation, see figure 1.2, where they had to accurately measure the angles in each of the many hundreds of triangles into which they had subdivided the territory to be surveyed. The surveyors emphasized the decimal aspect of the new Metric System by discarding the traditional (Babylonian system where angles are sub-divided into sixtieth parts) degrees and minutes of angular measurement, and instead divided the

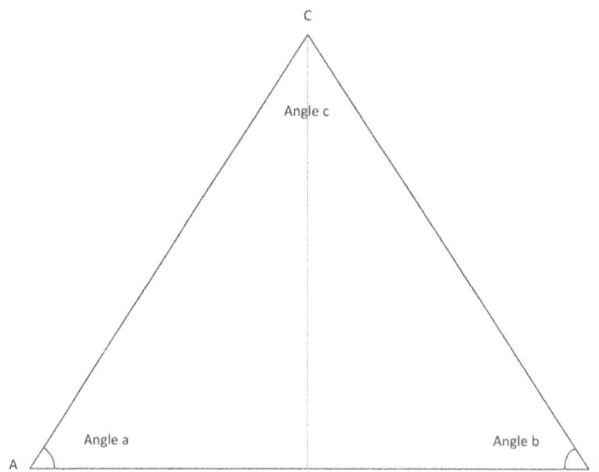

Figure 1.2. Triangulation: the known positions of two points, A and B, can be used to determine the relative position of a nearby point, C. The calculation involves the distance between the two known points, and the angles between the known points and the un-known point (as measured by the Borda circle). First, angle c is determined (we know angles a and b in triangle ABC), then AC and BC may be calculated by the Law of Sines. The altitude of the triangle or the offset of C from the horizontal line AB is found from AC.sin(a) or BC.sin(b); the offset in the opposite direction could be determined by using the cosines of a and b.

quadrant into one hundred grads that were then sub-divided into one thousand arc-minutes, that were in turn sub-subdivided decimally into arc-seconds.

At the completion of the measurements of each of the angles, in each of the triangles that made up the distance between Dunkirk and Barcelona, the surveyors had to measure as precisely as possible the length of one of the sides of one of these triangles. Then using trigonometry, they could calculate all of the distances, in all of the triangles. Consequently, two of the most important measurements were made by repeatedly laying a set of two-toise (a toise was a distance of about two yards—a fathom in English) platinum rulers mounted end to end on large wooden stabilizing blocks along two straight, flat roads.

In preparation for this final calculation of the metre, in September 1795 Méchain found a suitably flat stretch of road near Perpignan, the length of which he measured precisely with his platinum toise measuring sticks. This flat stretch of road was found from measurements made over several weeks by precisely placing the measuring sticks one in front of the other, to be a little over 6000 toise. This distance was one of the sides of one of the triangles being surveyed, and it would serve as a test for the quality of the angle measurements.

This final test of the quality of the surveying, and hence of the precision with which the metre had been determined, came in 1798 when Delambre took almost two months to measure the length of one of the sides of one of the surveyed triangles in the countryside near Paris. Again, the precise measurement of length was made with a pair of two-toise platinum measuring sticks. After this measurement of distance was completed in September 1798, the two surveyors met to compare their two length measurements. From the measured distance in the countryside near Paris, 6075.90 toise, they calculated a theoretical value for the length of the side of the triangle near Perpignan, which they calculated to be 6006.198 toise. Méchain then revealed that he had actually measured this distance to be 6006.27 toise. As the surveyors commented, 'the difference is negligible.'

But even after almost seven years of field measurements made under the most arduous of conditions, the astronomers' carefully calculated metre turned out to be no more precise than the preliminary estimate of the *Académie des sciences*, which the French authorities, to the annoyance of many *savants* (including the two surveyors), had promoted in the interim as a provisional unit. Sadly, because of the political situation in France and the physical and financial hardship of the surveying process, the two surveyors and their teams took a lot longer to complete their task than had originally been expected. As time went on, the politicians in Paris realized that to satisfy the demands of the people for a new system of weights and measures they could not wait for the precise measurements of Delambre and Méchain, so they adopted a provisional metre based upon the survey of the same meridian carried out by César-Francois Cassini, and made and distributed wooden and steel 'provisional metre-sticks'.

Today, we think nothing of making measurements of distance over extended sections of the globe. We have a network of satellites in orbit around the Earth, and this investment has made global positioning something that you have in your car when you are going out for dinner. But what of the measurements of those two late-18th Century surveyors, Delambre and Méchain, who set out into a war zone with the very latest of surveying equipment and measured the distance from Dunkirk to Barcelona with the greatest of precision? The angle measuring repeating circles, which had been designed by Jean-Charles de Borda and used by the two surveyors to measure the meridian (see figure 1.1) were an enormous improvement over previous surveying instruments. In themselves, the circles were very stable by virtue of being massive and large (made of dense metals); however, the problem was that they had usually to be mounted well above ground level to make a measurement. And consequently, the instability of the wooden platforms that the precision repeating circles were mounted upon were the main source of the error in the measurement; particularly, when the weather was stormy and windy.

To gain an idea of the precision of the metric survey, consider some data measured by Delambre near Dunkirk. Here, he needed to establish the latitude of his observations, and to do this he made detailed measurements of the transits of various well-known stars. Delambre made dozens of observations, which gave him a latitude of $51° 2'16.66''$; a value which changed by a fraction of an arc-second when he removed what he perceived to be the least-reliable measurement. Let us therefore assume a precision of 0.1 arc-second; to relate this estimate of measurement uncertainty to a distance, consider the circumference of the perfectly spherical Earth as being 40 000 kilometres. By dividing this circumference by 360, we determine the distance equivalent to measurement of a single degree, then by dividing by sixty and then again by sixty we determine the distance equivalent to a single arc-second; which gives an uncertainty in their measurements of about 3 m in the measurement of distance.

That is, these late-18th Century surveyors were capable of determining the location of an object on the surface of the Earth to a precision of about fourteen feet; provided they had sufficient time. But this remains an amazing feat when one considers that a commercial GPS system is only accurate to about three-times this level of precision; and the GPS represents an enormous investment, in terms of money and time in space-age technology.

The final results of the survey not only confirmed the value of the provisional metre established in 1793, but also produced something that was not anticipated; genuine new science, which we will explore in the coming chapters. Delambre and Méchain had not set out to do basic research, but merely to improve the precision of something that had previously been measured. But what they discovered was that the Earth was even less spherical than Sir Isaac Newton had calculated. According to surveying data determined by French *savants* in the mid-18th Century in Peru and France, the eccentricity of the Earth was roughly 1/300; that is, the Earth's radius to the Poles was 1/300 or 0.3% shorter than a radius to the Equator. Delambre and Méchain had found that this eccentricity (from Dunkirk to Barcelona) was about 1/150, or twice as great as had previously been thought.

The difference between the theory of Newton and the measurements of the surveyors had revealed that the Earth's surface was a patchwork of coupled segments—it was not uniform. As Méchain commented to a colleague, 'Our observations show that the Earth's curve is nearly circular from Dunkirk to Paris, more elliptical from Paris to Evaux, even more elliptical from Evaux to Carcassonne, then returns to the prior ellipticity from Carcassonne to Barcelona, So why did He who moulded our globe with his hands not take more care ... ?'; a splendid question, which modern geology can now answer.

As in the scientific advances of the late-19th Century, when scientists attempted to measure something familiar merely with greater precision, they invariably discovered new science; so with the metric survey. In the late-19th Century, increased precision in measurement led to the creation of quantum mechanics and relativity, in the late-18th Century, increased precision of measurement allowed the creation of 19th Century science.

1.4 Other surveys

The European Enlightenment had come up with the idea of constructing a new decimal metrology based on a single measurement of length. Such ideas, however, have a long history, and it is to Ancient China that we must turn for the first consistent use of decimal weights and measures; particularly, in the decrees of the first emperor, Chin Shih Huang Ti in 221 BCE.

Given the size of China, it is perhaps not surprising that an early effort was also made in fixing terrestrial length measurements in terms of astronomical measurements or observations. It was an early idea of Chinese *savants*, going back before the time of Confucius (551–479 BCE), that the shadow-length of a standard height (an 8 foot gnomon), at the summer solstice increased by 1 inch for every thousand *li* (a length measurement equivalent to 1500 *chi* or Chinese feet) north of the Earth's 'centre', and decreased by the same proportion as one went south. This rule of thumb remained current until the Han Dynasty (205 BCE–220), when detailed surveying of the expanding Chinese Empire showed it to be incorrect. But it was not until the Tang Dynasty (618–907) that a systematic effort was made to determine a range of latitudes. This extensive Tang survey had the objective of correlating the lengths of terrestrial and celestial measures by finding the number of *li* that corresponded to 1° of polar altitude; that is, terrestrial latitude; thereby fixing the length of the *li* in terms of the Earth's circumference. This Chinese meridian survey takes its place in history between the lines of Eratosthenes (c 200 BCE), and those of the astronomers of the Caliph, al-Ma'mūm (c 827), but more than a thousand years before the metric survey.

The majority of these Chinese surveying measurements were undertaken between 723 and 726 by the Astronomer-Royal, Nankung Yüeh, and his assistant, I-Hsing, a Buddhist monk. The survey was carried out at eleven sites along a meridian running from the Great Wall in the north to Indo-China in the south; a distance of 7973 *li* or about 2500 km. The main result of this field work was that the difference in shadow-length was found to be close to 4 in for each 1000 *li* north and south, and that the terrestrial distance corresponding to 1° of polar altitude was calculated to be 351 *li* and 80 *bu* (the *bu* was a measure of between 5–6 *chi*). The imperial surveyors had achieved their goal of defining a terrestrial unit of length, intended for use throughout the empire, in terms of the dimensions of 'Heaven and Earth'; that is, 1/351 of a degree.

This survey is today practically unknown, yet it represents an outstanding achievement given the spaciousness and amplitude of its plan and organization, and one of the earliest uses of advanced mathematics which was needed to compute the final result. These results were known in 18th Century Europe, as they were commented upon by Leonard Euler and later by Pierre Simon de Laplace. While the metric survey obtained a routine precision of about 1 part in 10^6 in distance, the much earlier Chinese survey could boast only of a precision of 1 part in 10^3. The Tang value of the *li* gives a modern equivalence of 323 m, but the earlier standard Han *li* is very different at 416 m.

Further reading

For further information about all aspects of the history of the Metric System mentioned in this chapter, please see Williams J H 2014 *Defining and Measuring Nature: The make of all things* (San Rafael, CA: Morgan & Claypool).

In addition, there are many useful entries in the online encyclopaedia, Wikipedia, about the topics touched upon in this chapter. And the best place to find further details about science in historical China is Needham J 1962 *Science and Civilization in China* (Cambridge: Cambridge University Press); the survey mentioned in this chapter is taken from volume 4 (part 1), pages 42–55

Chapter 2

The error in all things

2.1 Introduction

In 1792, two internationally renowned French mathematicians and astronomers set out from Paris; one headed north, the other south to measure the distance between Dunkirk and Barcelona with as much precision as was afforded by their technology. These two *géomètres* were seeking to lay the groundwork for a new universal system of measurement, by creating a new standard unit of length to be derived from nature herself. Forget the distance from the tip of the king's nose to the tip of his thumb, the *aune* and the ell, the confusing patchwork of local measures in *ancien régime* France, which only served to institutionalize fraud. The new standard of measurement, as befitted the spirit of the age of universal rights and ideals, would be based on something *universal*: the size of the Earth. The result of these endeavours would be a system of weights and measures appropriate for all people; irrespective of where they lived, or under what type of government.

Making long-distance measurements of great precision and accuracy is never easy. Engineers and geodesists today would use the Global Positioning System or laser range-finding theodolites to do this sort of work. Best practice in the late-18th Century involved fitting together a number of different kinds of observations, each demanding work in the field, or on top of a mountain with large, heavy instruments of brass and glass (see figure 1.1). Astronomical observations fixed the terrestrial coordinates of the endpoints of the line to be measured, but these points then had to be squared with a set of observations made along the line itself: first, the angles of triangles sighted from hilltop to hilltop along the entire distance of interest; second, the actual paced-off length of one side of one of those triangles, a length which could then be projected through the whole chain of triangles using trigonometry. It was an immensely complex business, demanding patience, fortitude, good physical health, fine eye-sight and a ready hand for kilometres of long arithmetic calculations.

Sadly, the French Revolution and the wars triggered by that event did not facilitate such an enterprise. As Méchain and Delambre set about their task, the countries around France declared war on France, which was a particular problem for Méchain's surveying team working in Spain. Indeed, the saga of the surveying process is one of the great epics of science. The surveying team would arrive in a small town and present their ormolu commission papers to surly peasant mobs led by sly ambitious local politicians, only to discover that the government who had issued these papers has since dissolved, and the politicians who had signed the papers with a great flourish had already been executed as counter-revolutionaries.

Not only did the surveyors have to deal with local peasants who had not the slightest interest in what they were doing or who were of the opinion that these educated gentlemen from Paris were themselves counter-revolutionary agents, but the measurements were not easy to make. Often it was necessary to construct a platform around the spire of a local church or on top of a hill so as to be able to see the next hill, or castle, or mountain in the chain of surveying points. Then, making the line of sight measurements of the angles between one point and another point became difficult because of the weather; storms, snow, mist and rain all contributed to the difficulty of the enterprise. In addition, the French currency became worthless through hyper-inflation and so the surveyors had to beg and steal food, they fell ill and Méchain almost died in a fall from an observation platform. All of these hardships contributed to the quality of the measurements they were making.

I am sure that any of my readers who have spent a longish period of time trying to make a measurement of something that is not easy to measure, will agree with me that there were good days in the laboratory, and then there were bad days in the laboratory. That is, on some days of measurements, you knew everything was working well, and that you were gathering 'good' data, but on another day something will not have been working so well and the data will not have seemed so consistent, and you will know that these newer data are less reliable. But how (and why) do you distinguish between the, perhaps, subjective good data and the less-good data? How do you sort and characterize your data? And if you have been making measurements over seven years, you will have a lot of data to sort. This was the other great problem for the surveyors; how to qualify and then quantify their measurements? No experimenters had previously gathered such a quantity of internationally important data, and European and American *savants* were waiting for these results. Delambre and Méchain were the first to undertake such an extended exercise; these two astronomer mathematicians were the first true scientists, the first to make high-precision measurements of anything.

Both surveyors survived the seven-years of field work, dodging war zones and the guillotine. They brought back data that enabled an international committee of mathematicians and geographers (assembled from those countries with which Frances was not at war—there were not many) to arrive at a value for the quarter meridian, and from it to derive the length of what would be called the *mètre*. A length that was for almost two centuries defined by a flat bar of platinum-iridium alloy, and was the basis of the metric system of weights and measures. Today, however, this length is defined by the speed of light (299 792 458 metres per second) and the definition of the second. Yet when the data brought back to Paris by the two surveyors was analyzed, the shape of the Earth was discovered to be a lot less regular than had been imagined. No elegant geometrical curve could accommodate the survey's data. Several centuries of scientific debate and theorizing about the form of the globe (was it a sphere, an oblate ellipsoid or something egg-shaped ... and if so, why?) was ending in a tautology: the shape of the Earth was the shape of the Earth. A unique, irregular, lop-sided thing. The seven-years of field work transformed the world by giving to science a new morphology for the globe; that irregular spheroid today called the geoid.

2.2 Méchain's 'error' in greater detail and least-squares

Every so often, much is made of a supposed error in the data measured by Méchain; an error that was, supposedly, kept a secret. However, the truth is that this secret is what the French call *le secret de polichinelle*; that is, not a secret at all, but something everyone knows. No one was seeking to hide anything.

Was there a hidden error that corrupted the pure, philosophical metre? Is that metre-stick wrong because of some French double-dealing? Of course, not. Admittedly, your metre is wrong, in that the distance from the North Pole to the Equator via Paris and Barcelona is more than 10 000 000 metres. And yes, there was some double-dealing in the observations used to calculate the metre's length; Méchain, seeking to fix the southern endpoint of his line, ran into some discrepancies in the results he calculated from his observations, and rather than report them directly, he fussed and fudged for years while trying to clean up the data. Anxiety over the whole affair, pushed Méchain to the edge of madness, and hounded him to his death in 1804.

But Méchain's fussing and fudging had only a minuscule effect on the length of the metre itself, a distortion completely lost in the mathematics that were required once it became clear that the Earth has an irregular form. So, what then is one ten-millionth of a quarter meridian? This, of course, depends on where you are and on what you want to measure.

How then, could Méchain's 'error', which came to light after his death, possibly have any influence or be of any importance? In *The Measure of All Things*, Ken Alder comments that the world was changed by the 'error itself'. As an experienced astronomer who is credited with the discovery of many stars and comets, Méchain could not bring himself to accept that he had made an error, but in fact, unbeknownst to him, in the complex set of interconnected measurements he had accumulated over seven years, among the many thousands of individual observations, which constituted his data he had caught a first glimpse of the phenomenon of error and of uncertainty. There was no mistake that he could find and simply correct. As he spent years fretting over his data, and losing his physical and mental health in the process, he was merely chasing a spectre. And the ghost he kept almost seeing as he stared at the columns of numbers was nothing less than the limits of his ability, or the ability of any measurement scientist to get the 'right answer' to the problem they are investigating.

After Méchain's death, when his colleagues re-worked his thousands of calculations and observations, they did not find an error but the concept of measurement uncertainty, and, in the early years of the 19th Century, they created a theory to deal with it, a metric of metrics. Science itself was in this way transformed. The plight of Pierre Méchain will make sense to any scientist familiar with the standard laboratory tools of error analysis, those mathematical techniques that enable researchers to separate precision from accuracy, and to assess what is attainable in a specific measurement. Méchain, however, had no such tools to hand.

Science would not be the same after the metric survey; left behind was the world of *savants* like Méchain and his co-surveyor for whom the pursuit of enlightenment was a simple confrontation with the demon of error personifying a lack of enlightenment and rationality. For them, error was like sin, and to resist temptation they bound their spirits to the rectitude of their measurements. But the metric survey transcended any previous collection of observations. A new form of science arose, and with it the new man, the scientist. That is, someone more functionary than wise sage or *savant*, a professional who calibrated himself along with his instruments. Gone were the high-priests of Reason, who had been the souls of their measuring devices. Modern science was born, along with the idea of an inherent unavoidable uncertainty in measurement.

2.3 The metric survey

The geodetic survey done in Peru in the early-18th Century by a group of French scientists serves as a good example of how such surveys were conducted at that time. The survey consisted of two teams, who perform the same measurements and calculations, and then compared their results. In Peru, the French were trying to measure the length of a degree of arc near the equator and, using a French length unit, the toise which is 6 Paris feet or 6.39 English feet, one team got a result for a degree of longitude of 56 749 toise and the other team's result was 56 768 toise.[1]

Pierre Méchain was a meticulous astronomer, but given to obsessive attention to detail. His own measurements did not agree exactly nor meet his exacting standards, though they are, in retrospect, some of his best. He ended up fudging results, changing figures to make himself look better or no worse than his co-surveyor, but in reality he was trying to cover up errors that were, to him, intolerable. When Delambre received Méchain's raw data, after Méchain had died of yellow fever while trying to correct his observations by surveying further south than Barcelona, the data was not in bound notebooks but on scraps of paper with erasures, lack of dates, etc (see figure 2.1). Delambre carried forward his colleague's cover-up, cleaning up the data to make it publishable. He did this because he found that Méchain's final results were correct, erring mostly in the size of variations among his observations. The erasures and corrections did not affect the final result, but made Méchain's work look as if it had been better performed than it actually had been. Various instrumental problems, such as excessive wear and lack of calibration, were also observed.

Twenty-five years after Méchain's death, a young astronomer named Jean-Nicolas Nicollet (1786–1843) resolved the enigma of the 'error'. The problem was

---

[1] This is a difference of 19 toise, or 0.0335%. That's about 120 (English) feet difference over about 68.73 miles; again a remarkably precise result given the period and the locality.

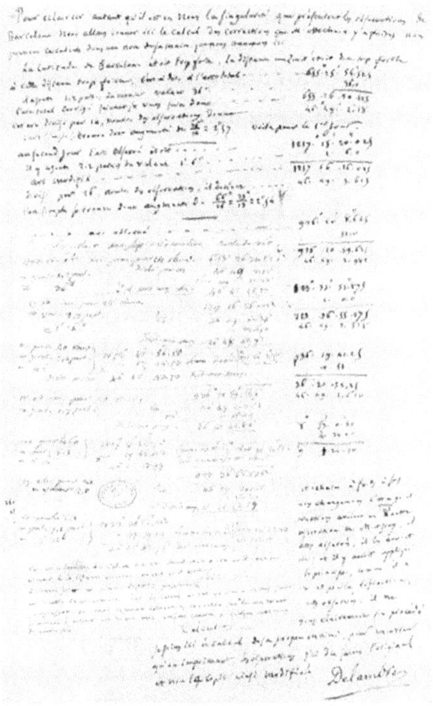

Figure 2.1. Méchain's data with notes by Delambre. Between 1806 and 1810, Delambre reconstructed Méchain's logbook by pasting into a bound register the loose sheets on which Méchain had recorded his data. Delambre organized the sheets into chronological order, retraced Méchain's pencilled data in ink, and indicated the provenance of each document (that is, Delambre followed what is still sound practice). On this particular page Delambre has pasted Méchain's observations from Barcelona for 15 December 1793. In the margin Delambre notes: '*Here are some changes that Méchain has made to the angle measurements for which it is difficult to imagine a legitimate rationale*'. He goes on to explain that Méchain's calculations on this page leave no doubt whatsoever that the corrections are not legitimate, but serve only to make the data appear more precise than they actually are. (From the *Archives de l'Observatoire de Paris*).

that Méchain and his contemporaries did not make a principled distinction between precision (that is, the internal consistency of results) and accuracy (that is, the degree to which those results approached the 'right answer'—whatever that might be).

The definitions for precision and accuracy given above are the key. Look at the Peru survey, where there was a difference of 19 toise between two results, this is a situation that occurs regularly in sequences of measurements. The second time you measure something the result may not agree with the first time it was measured, or other subsequent measurements. What is causing the problem? If you are surveying in the Andes, could it have been the weather? It was warmer on day two than day one, so maybe the metal tape measure or ruler had expanded with the heat, or was the measurement more difficult to make on the second day, because you were not feeling well, or there had been an avalanche and the flat surface you needed to measure was not as flat as before. These are all sources of error, and they would perturb your measurement. Maybe on day one you did the measurement, and on day two you had a colleague repeat the measurement. There can be individual human variations in the way that measurements are set up, calibrated and used. So how can anything be measured accurately and with precision?

The method that has been developed in the last 200 years to analyze or to quantify such perturbations of a measurement is called statistical analysis, and includes techniques and concepts for dealing with multiple measurements. It is often the case that a detailed analysis of sets of measurements can require a detailed statistical analysis, which can be a lot less interesting than doing the measurements themselves, and this was probably what caused Lord Ernest Rutherford to make his famous comment, 'If you need statistics to analyse your experimental data, you should have done a better experiment'. But in modern science, it is not possible to avoid statistics, and it is a requirement for good practice (and it is demanded by the editors of journals, who decide when error is too great to bear).

It was subsequently discovered that in no instance had Méchain's alterations distorted the final result by more than two arc-seconds, meaning that his adjustments were minor compared to the uncertainties caused by the observer's inability to correct entirely for the refraction of light in the Earth's atmosphere. He had edited his results, not to alter the final outcome, but to keep up appearances as an astronomer. Delambre wrote, 'Undoubtedly Méchain was wrong not to publish these observations as he found them, and to modify them in such a way as to make them appear more precise and consistent than they were. But he always chose his final values in such a way as to ensure that the average was not altered, so there was no real harm in his action, except for the fact that another observer who published unadulterated numbers would be judged less capable and careful'. This is almost a definition of art, rather than the application of scientific principles. What was needed was a sound set of rules upon which an experimentalist could base his choice of which of his measurements they felt to be the least, or the most reliable.

On the subject of Méchain's 'error', Delambre preferred to blame the Earth, rather than his colleague. As he pointed out, the meridian project had confirmed that the shape of the Earth was irregular, and that not all meridians were equal. Delambre hypothesized that Méchain's readings had been distorted by local irregularities in the Earth's crust or by nearby mountains. This concern was not new, as Isaac Newton had tried to estimate the gravitational pull of mountains. But all of this was speculation as 18th Century technology did not permit measurements sufficiently precise to determine the effect of mountains on the value of the local acceleration due to gravity. Today, however, the science of geodesy consists principally of mapping these gravitational effects. Physicists and engineers estimate the pull of mountains on rockets and missiles. Some of the maps that chart the contours of the geoid are classified as military secrets.

Pierre Méchain and his contemporaries did not make a principled distinction between precision and accuracy. The two are not the same; precise results may appear 'reliable' in the sense that they give very nearly the same answer when measured again and again; yet they may lack validity in that they deviate consistently from the 'right answer'. Of course, in practice, distinguishing between the two can be extremely difficult as the 'right answer' is unknown.

Repetition of measurements using the Borda circle was designed to improve precision by reducing those errors that stemmed from the imperfect senses of the observer or the imperfect construction of the instrument's gauge (the sort of errors we would today characterize as falling into a random distribution). The Borda circle, however, was still subject to errors caused by the design of the instrument; the sort of errors we would today characterize as those constant (or systematic) errors which make results inaccurate, whatever their level of precision. Constant errors generally go undetected, as long as they stay constant. And in an intuitive way, Méchain and Delambre like their contemporaries understood this, hence their vigilance about maintaining a consistent setup for their apparatus from one series of observations to the next. What they failed to appreciate, however, was that the same repetition that enhanced precision might reduce accuracy. For instance, constant manipulation of the circle might wear down the instrument's central axis and, over time, cause the circle to tilt ever so slightly from the perpendicular. It was this unanticipated drift in the constant error, Nicollet suggested, that was the source of Méchain's discrepancies. Without a concept of error to help him identify the source of this contradiction, Méchain was confounded.

Oddly enough, Nicollet noted, it was Méchain's own obsessiveness which made it possible to confirm the cause of the discrepancy, and to correct for it. One needed to compensate for any change in the instrument's verticality by balancing the data for stars which passed north of the zenith (the highest point of the midnight sky) against those which passed south of it. Because Méchain had measured so many extra stars, such an operation was possible from the recorded data.

To calculate the latitude, Méchain had first calculated the average latitude implied by each star he measured, and then averaged all the averages, giving equal weight to each. Nothing could be simpler (or more naïve). Nicollet, by contrast, first analyzed the data for the stars Méchain had measured that passed north of the zenith, taking the average of the average latitude implied by each. Then Nicollet separately did the same thing for the stars Méchain had measured to the south of the zenith. Clustered in this manner, the results seemed to lack precision; at Barcelona the average latitude implied by the north-going stars differed from the average latitude implied by the south-going stars by 1.5 arc-seconds. At Fontana de Oro they differed by 4.2 arc-seconds. But when the northern average and the southern average were themselves combined at each location, they suggested a remarkable accuracy; the combined latitude for the Fontana de Oro agreed with the combined latitude from Barcelona to within 0.25 arc-seconds. Nicollet had demonstrated that there was no discrepancy, and that Méchain's reported value for Barcelona was within 0.4 arc-seconds of the answer indicated by his data; when properly analysed after a post-mortem of Méchain's surveying instrument.

2.4 Least-squares

Hindsight and time are ideal for solving problems that were originally thought to be mysteries. But the confusions about the results of the metric survey did start *savants* thinking about the nature of the 'right answer', and about the possible existence of the 'correct value' of a natural phenomenon. It may be more of a theological question to ask if perfection exists in nature awaiting discovery, but Delambre was neither a believer nor an atheist. He was a sceptic, for whom perfect knowledge lay beyond man's grasp; so why should anyone expect him to produce a perfect metre?

In coming to terms with the imperfections and limits of the experimenter's art, Delambre had a powerful new tool at his disposal, one which he and Méchain had inadvertently inspired, but which he alone had lived to see. Previously, *savants* had sought to fit imperfect data to a perfect curve. Astronomers had agreed that the Earth was an oblate ellipsoid, but they had been unable to agree on its degree of eccentricity, which now seemed, moreover, to vary from place to place (the same would also have been the case for the Chinese survey of the 8th Century mentioned in chapter 1). Assume that the data had been gathered by fallible (but exacting) investigators using fallible (but ingenious) instruments on a (possibly) lumpy irregular Earth, and then ask yourself: what was the best curve through the data, and how much did the data deviate from that curve? This was the question asked by the mathematician Adrien-Marie Legendre (1752–1833).

A contemporary of Laplace and Delambre, Legendre was elected to the *Académie des sciences* at the age of thirty. In 1788 he showed the geodesists on the Paris–Greenwich surveying expedition how to correct for the curvature of their triangles. Appointed with Cassini and Méchain to the metric survey, he withdrew in favour of Delambre. Later he was one of the *savants* who calculated the length of the metre for the international commission who received the data of the surveyors. He was as baffled by the result (an unexpected eccentricity of 1/150) as were the other members of the commission. Legendre's answer, the method of least-squares, has since become a standard tool of statistical analysis. It was also among the most important breakthroughs in mathematical science; not because it produced new knowledge of nature, but because it produced a new way of conceptualizing and quantifying error.

For centuries *savants* had felt entitled to use their intuition and experience to publish their single 'best' observation as the measure of a phenomenon. During the course of the 18th Century, *savants* had increasingly come to believe that the arithmetic mean of their measurements offered the most balanced view of their results. Yet many *savants* continued to feel, like Méchain, that any measurement that strayed too far from the mean ought to count for less than those near to it, and hence could be discarded.

Adrien-Marie Legendre suggested a practical solution; that the best curve would be the one that minimized the square of the value of the difference of each data point from that theoretical curve. This was a general rule, and it was also a tractable calculation that could be widely understood. Legendre's least-squares method played off the intuition that the best result should strike a balance among divergent data, much as the centre of gravity defines the balance point of an object. As he noted, the least-squares method also justified choosing the arithmetic mean in the simplest of cases. It gave *savants* a workable method for weighting and sorting data.

In 1805, just as Delambre was completing the first volume of his biography of the metric survey, Legendre tried out his method on what was now the world's most famous data set, the one he had puzzled over ever since Delambre and Méchain had handed it to the international commission in 1798. Legendre assumed that the Earth's meridian traced out an ellipse; he then used the least-squares rule to find the eccentricity that would minimize the square of each latitude's deviation from that curve as it connected the data points. In this way, he observed that the deviations of the various latitudes from that optimal curve remained sufficiently large to be ascribed to the figure of the Earth and not to the data; it was the Earth that was difficult to understand, not the data.

Four years after Legendre's paper, the great mathematician Karl Friedrich Gauss (1777–1855) claimed that he had been using the least-squares rule, which he called 'my method', for nearly a decade. As often happens, this simultaneous discovery was no coincidence. Indeed, Gauss was working on the same data set, the data from the metric survey, which had been published in Germany in 1799. They were also reading the same mathematicians, especially Laplace. However, as so often happens, this simultaneous discovery prompted a bitter dispute over priority. In this instance, there seems little doubt that the two men arrived at the method independently, although it was Legendre who published first. But there is no doubt that it was Gauss who pointed out the method's important meaning.

Gauss had predicted the position of an asteroid that has been lost to observation while passing behind the Sun. The asteroid in question, Ceres, had been observed 24 times early in 1801 by Giuseppe Piazzi who published the positions. Gauss used the observations to calculate where and when the asteroid would re-appear. He was correct to within half a degree. Among the mathematical tools he used in this truly impressive calculation was the least-squares method, to match observations with a model developed from the observations assuming that the experimental or observational errors followed a normal distribution. He took the observations of others, and used statistics and probability theory to construct a theoretical model, which he used to predict where Ceres would be some time into the future; that is, to define the path and the velocity of the asteroid on the path. Gauss got it right, thereby demonstrating his own genius and the power of his technique, as nothing proves the value of new mathematical models like predictions that are validated by subsequent experiment. Thus was modern physical science born.

Adrien-Marie Legendre presented his method of least-squares as workable and plausible. Gauss justified it by showing that it gave the most probable value in those situations where the errors were distributed along a 'bell curve' (known today as a normal or Gaussian distribution), see figure 2.2. This probability-based approach enabled Laplace to show, in 1810–11, that the least-squares method had the following advantages: it best reduced the error as the number of observations increased; it indicated how to distinguish between random errors (which define precision) and constant errors (which define accuracy); and it suggested how likely it was that the chosen curve was the best curve. In their search for an illusory perfection, *savants* had learned not only how to distinguish between different kinds of error, but also that error could be approached with quantitative confidence. The years between 1805 and 1811 saw the rise of a new scientific theory; not a theory of nature, but a theory of error. It was this theory that would allow Nicollet to redeem

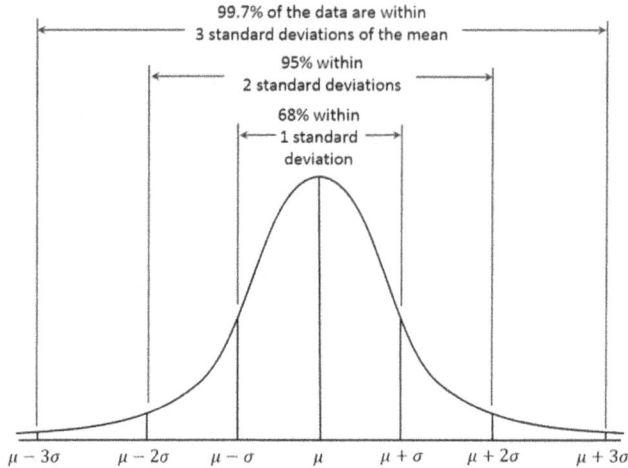

Figure 2.2. Curve representing a Gaussian or Normal 'Bell' distribution, where the mean of the distribution is μ and the standard deviation is σ. For such a normal distribution, the values less than one standard deviation away from the mean account for 68.27% of the set; while two standard deviations from the mean account for 95.45%; and three standard deviations account for 99.73% (this is the integrated area under the curve).

Méchain's reputation by distinguishing between those errors that were random and those that were systematic.

During the course of the next century science learned to manage uncertainty, to look at measurements as having an intrinsic uncertainty. The fields of statistics and probability that would one day emerge from the insights of Legendre, Laplace, and Gauss would transform the physical sciences, inspire the biological sciences, and give birth to the social sciences. In the process, *savants* became scientists.

Men like Delambre, Laplace, and Legendre began as *savants*, but they were changed by the struggle to quantify the uncertainty that always exists in studying nature. They would be some of the first non-philosophers and non-theologians to ask: how confident are we that we know what we think we know? They sought to rid themselves of value judgments about nature, and to perform measurements with detachment. Pierre Méchain lived and died a *savant*. Measurement mattered to him as much as it did to the *ancien régime* peasants, bakers, and ordinary families. Whether it was the location of a star or the weight of a loaf of bread, measurement expressed value. Measuring something was a moral act. For the *savant*, the pattern of the heavens revealed an aspect of an overarching structure. To measure the shape of the Earth or the position of a star was to fix its place in that overall pattern of things, just as the weight of a loaf of bread sustained the just price for bread and helped to maintain a social order.

In 1792, the extreme radical Jean-Paul Marat had been the first person to tag *savants* with the name *scientifiques* when he referred sneeringly to the academicians' self-serving project to measure the Earth in order to create uniform weights and measures. But the name stuck, even though we have continually to justify our existence.

2.5 Statistical methods

One important concept in statistics is the mean of a data set: commonly referred to as an average. For the measurements made by the Peruvian survey mentioned above, there were two measurements of a distance, one from each surveying team These two distances, x_1 and x_2 would have then been added together and divided by the number of measurements to derive the mean. But more information is provided by indicating the range of the two measurements. The two measurements in question were 56 749 toise and 56 768 toise; the average being 56 758.5 toise. But these uncertainties in the measurement could also be written as 56 749 ± 9.5 toise and 56 768 ± 9.5 toise; the amount that may be added or subtracted from the measured value indicates the range of values that could be expected if the measurement were repeated again. But such an approach to data analysis was not able to help Méchain in analysing his data.

Subsequent to his re-evaluation of Méchain's data, Nicollet used the same kinds of statistical methods as had Méchain to deal with the discrepancies in Méchain's observations, and was able to show that these discrepancies were not errors, but variations within the tolerable for a precise and accurate understanding of the material needed to establish the correct length of the metre. Statistics are not measurements, but a formulation that permits the building of models from measurements. There are many statistical methods available for the analysis of data, which allow scientists to identify and quantify the unavoidable variance or variation of measurements. If these methods are used correctly, others may dispute some parts of the questions being addressed, but not how the conclusions were reached through the statistical analysis. If one starts with an assumption, proceeds to gather data about that assumption, develops an hypothesis, then performs a statistical analysis on the data to see if the initial assumption is sound, any questions that may be asked are about the initial assumption, and not about the statistical procedure.

The method of least-squares is today a standard technique in regression analysis for the approximate solution of overdetermined systems; that is, to sets of equations in which there are more equations than there are unknowns. The name signifies that the overall solution minimizes the sum of the squares of the errors made in the results of every single equation or data point. The best fit in the least-squares sense minimizes the sum of squared residuals; a residual being the difference between an observed value and the fitted value provided by a model.

Least-squares problems fall into two categories: linear or ordinary least-squares and non-linear least-squares, depending on whether or not the residuals are linear in all unknowns. The linear least-squares problem occurs in statistical regression analysis; it has a closed-form solution. The non-linear problem is usually solved by iterative refinement; at each iteration the system is approximated by a linear one, and thus the core calculation is similar in both cases.

The method of least-squares grew out of the fields of astronomy and geodesy as astronomers and mathematicians sought to provide solutions to the challenges of navigation, as the accurate description of the behaviour of celestial bodies was the key to enabling ships to navigate in the open sea. The method was the culmination of several advances that took place during the course of the 18th Century. Firstly by *savants* attempting to combine numerous observations taken under the same conditions, as opposed to simply trying ones best to observe and record a single observation as accurately as possible. The approach was known as the method of averages, and was notably used by the German astronomer Tobias Mayer (1723–62) while studying the librations of the Moon in 1750, and by Pierre Simon de Laplace in his work in explaining the differences in motion of Jupiter and Saturn in 1788.

Then *savants* tried to combine observations made under different conditions. The method came to be known as the method of least absolute deviation. It was derived by the polymath, poet and priest Roger Joseph Boscovich (1711–87) in his work on the shape of the Earth in 1757, and by Pierre Simon de Laplace working on the same problem in 1799. And finally, there was the concept of probability and the development of criteria that could be evaluated numerically to determine when the solution with the minimum error has been achieved. Laplace tried to specify a mathematical form of the probability density for errors and define a method of estimation that minimized the error of estimation. For this purpose, he used a symmetric two-sided exponential distribution we now call the Laplace distribution to model the error distribution, and used the sum of absolute deviation as the error of estimation.

As we have seen, the first clear and concise presentation of what today we call the method of least-squares was made by Legendre in 1805. The technique can be described as an algebraic procedure for fitting linear equations to data, and Legendre demonstrated the new method by analyzing the same data that Laplace had examined, the data accumulated by Delambre and Méchain to define the shape of the Earth. The value of Legendre's method of least-squares was immediately recognized by leading astronomers and geodesists of the time.

After reading Gauss's successful mathematical modelling of the orbit of Ceres, Laplace was able to use the same methods to prove the central limit theorem, and used it to give a large sample justification for the method of least-squares analysis and the normal distribution. In 1822, Gauss was able to state that the least-squares approach to regression analysis is optimal in the sense that in a linear model where the errors have a mean of zero; that is, are uncorrelated and have equal variances, the best linear unbiased estimator of the coefficients is the least-squares estimator; this is the Gauss–Markov theorem.

The problem consists in adjusting the parameters of a model function to best fit a data set (that is, data fitting; which as we will see later is very much the basis of modern physical science). A simple data set consists of n points or data pairs (x_i, y_i), $i = 1, ..., n$, where x_i is an independent variable and y_i is a dependent variable whose value is found by observation. The model function has the form $f(x, \beta)$, where m adjustable parameters are held in the vector β. The goal is to find the parameter values for the model which best fits the data. The least-squares method finds its optimum when the sum, S, of squared residuals

$$S = \sum_{i=1}^{n} r_i^2$$

is a minimum. A residual being defined as the difference between the actual value of the dependent variable and the value predicted by the model. Each data point has one residual, r_i.

$$r_i = y_i - f(x_i, \boldsymbol{\beta}).$$

Both the sum and the mean of the residuals are equal to zero.

An example of a model would be a straight line in two dimensions. Denoting the y-intercept as β_0 and the slope as β_1, the model function may be written as $f(x, \boldsymbol{\beta}) = \beta_0 + \beta_1 x$.

A data point may consist of more than one independent variable; for example, fitting a plane to a set of measurements, the plane is a function of two independent variables. In the most general case, there may be one or more independent variables and one or more dependent variables at each data point.

The minimum of the sum of squares is found by setting the gradient to zero. Since the model contains m parameters, there are m gradient equations:

$$\frac{\partial S}{\partial \beta_j} = 2 \sum_i r_i \frac{\partial r_i}{\partial \beta_j} = 0, \quad j = 1, \ldots, m,$$

and as $r_i = y_i - f(x_i, \boldsymbol{\beta})$, the gradient equations become

$$-2 \sum_i r_i \frac{\partial f(x_i, \boldsymbol{\beta})}{\partial \beta_j} = 0, \quad j = 1, \ldots, m.$$

The gradient equations are the central element of all least-squares problems; each particular problem requires particular expressions for the model and its partial derivatives.

A fitting model (also called a regression model) is linear when the model comprises a linear combination of the parameters; that is,

$$f(x, \boldsymbol{\beta}) = \sum_{j=1}^{m} \beta_j \phi_j(x),$$

where the function φ_j is a function of x.

Letting

$$X_{ij} = \frac{\partial f(x_i, \boldsymbol{\beta})}{\partial \beta_j} = \phi_j(x_i),$$

we can see that here the least-square estimate (or estimator, in the context of a random sample), $\boldsymbol{\beta}$ is given by

$$\underline{\beta} = \left(X^T X\right)^{-1} X^T y.$$

There is no closed-form solution to a non-linear least squares problem. Instead, readily available numerical algorithms are used to find the value of the parameters β that minimizes the function of interest. Most algorithms involve choosing initial values for the various parameters. Then, the parameters are refined iteratively; that is, the values are obtained by successive approximations:

$$\beta_j^{k+1} = \beta_j^k + \Delta\beta_j,$$

where k is a number defining the order of the iteration, and the vector of increments $\Delta\beta_j$ is termed the shift vector. With some commonly used algorithms, at each stage of iteration the model may be linearized by approximation to a first-order Taylor series expansion about β^k:

$$f(x_i, \beta) = f^k(x_i, \beta) + \sum_j \frac{\partial f(x_i, \beta)}{\partial \beta_j}(\beta_j - \beta_j^k)$$

$$= f^k(x_i, \beta) + \sum_j J_{ij}\Delta\beta_j.$$

The Jacobian J is a function of constants, the independent variable and the parameters, so it changes from one iteration to the next. The residuals are given by

$$r_i = y_i - f^k(x_i, \beta) - \sum_{k=1}^{m} J_{ik}\Delta\beta_k = \Delta y_i - \sum_{j=1}^{m} J_{ij}\Delta\beta_j.$$

To minimize the sum of squares of r_i, the gradient equation is set to zero and solved for $\Delta\beta_j$:

$$-2\sum_{i=1}^{n} J_{ij}\left(\Delta y_i - \sum_{k=1}^{m} J_{ik}\Delta\beta_k\right) = 0,$$

which, on rearrangement, become m simultaneous linear equations:

$$\sum_{i=1}^{n}\sum_{k=1}^{m} J_{ij}J_{ik}\Delta\beta_k = \sum_{i=1}^{n} J_{ij}\Delta y_i \qquad (j = 1, \ldots, m).$$

These may be written in matrix notation as

$$(J^T J)\Delta\beta = J^T \Delta y,$$

which are the defining equations of the Gauss–Newton algorithm.

   Differences between linear and non-linear least-squares: The model function, f, in linear least-squares (LLSQ) is a linear combination of parameters of the form $f = X_{i1}\beta_1 + X_{i2}\beta_2 + \cdots$ The model may represent a straight line, a parabola or any other linear combination of functions. In non-linear least-squares (NLLSQ) the parameters appear as functions such as β^2 or $e^{\beta x}$. If the derivatives $\partial f/\partial \beta_j$ are either constant or depend only on the values of the independent variable, the model is linear in the parameters. Otherwise the model is non-linear. Some of the differences between LLSQ and NLLSQ are given below, but fuller details may be found in any standard textbook.

   • Algorithms for finding the solution to a NLLSQ problem require initial values for the parameters, LLSQ does not.

- Like LLSQ, solution algorithms for NLLSQ often require that the Jacobian be calculated. Analytical expressions for the partial derivatives can be complicated. If analytical expressions are impossible to obtain, either the partial derivatives must be calculated by numerical approximation or an estimate must be made of the Jacobian.
- In NLLSQ non-convergence (that is, failure of the algorithm to find a minimum) is a common phenomenon whereas the LLSQ is globally concave so non-convergence is not an issue.
- NLLSQ is usually an iterative process. The iterative process has to be terminated when some defined convergence criterion has been satisfied. LLSQ solutions can be computed using direct methods, although problems with large numbers of parameters are usually solved by iterative methods.
- In LLSQ the solution is unique, but in NLLSQ there may be multiple minima in the sum of squares.

These differences should be considered whenever a non-linear least squares problem is being investigated.

The method of least-squares is often used to generate estimators and other statistics in regression analysis. Consider a simple harmonic spring obeying Hooke's law, where the extension of a spring y is proportional to the applied force, F. This is the basis of the analysis of molecular vibrations in infrared spectroscopy, and of the dynamics of engineered structures. We can write

$$y = f(F, k) = kF,$$

which constitutes the model, where F is the independent variable. To estimate the force constant, k, a series of n measurements with different forces will generate a data set (F_i, y_i), $i = 1, \ldots, n$, where y_i is a measured spring extension. Each experimental observation will contain some error. If we denote this error ε, we may specify an empirical model for our observations,

$$y_i = kF_i + \varepsilon_i.$$

There are many methods that could be used to estimate the unknown parameter k. Noting that the n equations in the m variables in our data comprise an over-determined system with one unknown and n equations, we may choose to estimate k using least-squares. The sum of squares to be minimized is

$$S = \sum_{i=1}^{n} (y_i - kF_i)^2.$$

The least-squares estimate of the force constant, k, is given by

$$\hat{k} = \frac{\sum_i F_i y_i}{\sum_i F_i^2}.$$

Here it is assumed that application of the force causes the spring to expand and, having derived the force constant by least-squares fitting, the extension can be predicted from Hooke's law. This is how a great many, much more complex problems are investigated today (see chapter 9).

In such an analysis, the investigator specifies an empirical model; for example, a straight line model which is used to test if there is a linear relationship between dependent and independent variables. If a linear relationship is found to exist, the variables are said to be correlated. However, correlation does not prove causation, as both variables may be correlated with other, hidden, variables, or the dependent variable may 'reverse cause' the independent variables, or the variables may be otherwise spuriously correlated. For example, suppose a correlation is 'found' between the number of deaths by drowning and the volume of soft-drinks sold at a beach. Yet, both the number of people going swimming and the volume of drinks sold increase with the rising temperature, and presumably the number of deaths by drowning is correlated with the number of people going swimming. Perhaps it is an increase in the number of swimmers that causes both variables to increase.

In order to make statistical tests on a set of measurements it is necessary to make assumptions about the nature of the experimental errors. A common (but not necessary) assumption is that the errors can be described by a normal distribution. This assumption is supported by the central limit theorem in many cases. However, even if the errors are not normally distributed, a central limit-like theorem will often, nonetheless imply that the parameter estimates will be approximately normally distributed as long as the sample is reasonably large. So given that the error mean is independent of the independent variables, the distribution of the error term is not an important issue in regression analysis.

In a least-squares calculation with unit weights, or in linear regression, the variance on the jth parameter, denoted $\mathrm{var}(\hat{\beta}_j)$, is usually estimated with

$$\mathrm{var}\left(\hat{\beta}_j\right) = \sigma^2\left([X^TX]^{-1}\right)_{jj} \approx \frac{S}{n-m}\left([X^TX]^{-1}\right)_{jj},$$

where the true residual variance σ^2 is replaced by an estimate based on the minimized value of the sum of squares objective function S. The denominator, $n-m$, is the statistical degree of freedom.

Confidence limits can be found if the probability distribution of the parameters is known, or if an asymptotic approximation can be made, or assumed. Likewise statistical tests on the residuals can be made if the probability distribution of the residuals is known, or assumed. The probability distribution of any linear combination of the dependent variables can be derived if the probability distribution of experimental errors is known, or assumed. Inference is particularly straightforward if the errors are assumed to follow a normal distribution, which implies that the parameter estimates and residuals will also be normally distributed conditional on the values of the independent variables.

Further reading

Details of the mathematics of the least-squares method, and how it may be applied to a range of problems may be found on the Internet. The online encyclopaedia, Wikipedia, has several excellent entries for the statistical methods used in data analysis; particularly, the least-squares method. For specific applications of the methodology to individual problems, the best place to start would be to Google the specific application.

With regard to the details of the epic surveying adventure of Delambre and Méchain to define the length of the metre, a reader could do no better than read *The Measure of All Things* by Ken Adler, published by Abacus, 2004. This is a readable, non-scientific but detailed account of the metric survey. Even though the volume is long at 370 pages, with an additional 200 pages of notes and references, it is informative and easy to read, with some thought-provoking observations on the difficulties of undertaking detailed scientific work in two countries that are actually at war.

Chapter 3

A language for measurement

3.1 Introduction

In previous chapters, we saw how Pierre Méchain and his contemporaries did not, in their measurements, make a distinction between precision; that is, the consistency of results, and accuracy; that is, the degree to which these results approached the 'right answer.' The 'right answer' being the great unknown. The modern nomenclature of measurement science seeks to avoid the possible ambiguities that may arise through the use of terms such as accuracy and the 'right answer'.

Having to use a new, but established nomenclature to describe something with which you feel you are familiar, and which you describe adequately in your own manner, can often seem like being forced to use a new system of weights and measures, or even a new language. It is irritating and confusing, but strictures being imposed on how you describe science and technology are there for the best of reasons. That is, that we may be confident that we are all talking about the same thing, and in the same manner.

A measurement tells you about a property of something you are investigating, giving it a number and a unit; for example, x grams cm^{-3} or y kelvin. Measurements are made using an instrument of some kind; rulers, stop-clocks, chemical balances, thermometers, or even the Large Hadron Collider at CERN are all measuring instruments.

3.2 The quality of measurements

Evaluating the quality of a measurement is an essential step on the route to drawing sensible conclusions from that measurement, and, consequently, scientists have evolved a special vocabulary to help them think clearly about their data. Key terms that describe the quality of measurements are:

- validity;
- accuracy;

- precision (repeatability or reproducibility);
- measurement uncertainty.

3.2.1 Validity

A measurement is valid if it measures what it is supposed to be measuring. If a factor or a condition in a measurement is uncontrolled, the measurements may not be valid; for example, if you are investigating the heating effect (a measure of power, P) of an electric current, I, flowing through a wire with an electrical resistance, R, by increasing the current (that is, you are exploring the equation $P = I^2R$), the resistance of the wire may change as it is heated by the flowing current. This heating effect may well invalidate your measurements, as R is a function of temperature, and so your results (at a level of precision determined by the magnitude of the temperature dependence of R) could be skewed.

3.2.2 Accuracy

This describes how closely a measurement approaches the 'true value' of a physical quantity. The 'true' value of a measurement is the value that would be obtained by a perfect measurement; that is, in a perfect or ideal world. As the true value is not known, accuracy can only ever be a qualitative term. Many measured quantities have a range of values rather than one 'true' value. For example, a collection of electrical resistors all marked 1 kΩ will have a range of values because they are mass produced, but the mean value of a collection of such resistors should be close to 1 kΩ. The variation enables you to identify: a mean, a range and the distribution of values across the range.

3.2.3 Precision

The closeness of agreement between replicated measurements on the same or similar objects under specified conditions. We could also say that this is the extent to which a measurement replicated under the same conditions gives a consistent result.

3.2.4 Measurement uncertainty

The uncertainty of a measurement is the doubt that exists about its value. For any measurement, even the most carefully undertaken measurements, there is always a margin of uncertainty. (This will be the central topic of much of this volume.) The uncertainty about a measurement has two aspects:
- the width of the margin, or interval. This is the range of values within which one expects the true value to be found. (Note: this is not necessarily the range of values one might obtain when taking measurements of that property, which may include outliers.)
- confidence level; that is, how sure the experimenter is that the true value lies within that margin or interval.

Uncertainty in measurements can be reduced trivially by using an instrument that has a scale with smaller divisions. For example, if you use a ruler with a centimetre scale then the uncertainty in a measured length is likely to be, of order, plus and minus one centimetre. A ruler with a millimetre scale would reduce the uncertainty in length to, of order, plus and minus one millimetre. This was the idea behind the use of a decimal metric scale, as opposed to the traditional Babylonian angle divisions for the surveying instruments used in the metric survey (the Borda circles), see section 1.3.

3.3 Measurement errors

It is important not to confuse the terms 'error' and 'uncertainty'. Error refers to the difference between a measured value and the true value (which we do not know) of the physical quantity being measured. Whenever possible we try to correct for any known errors; for example, by applying corrections from calibrations. But any error whose value we do not know is a source of uncertainty. Measurement errors can arise from two sources: a random component, where repeating the measurement gives an unpredictably different result; and a systematic component, where the same influence affects the result for each of the repeated measurements.

Every time a measurement is made under what seem to be the same conditions, random effects can still influence the measured value. A series of measurements therefore produces a scatter of values about a mean value. The influence of variable factors may change with each measurement, changing the mean value. Increasing the number of observations generally reduces the uncertainty in the mean value. Systematic errors (measurements that are either consistently too large, or consistently too small) can result from:

- poor technique (for example, carelessness with parallax corrections when reading a scale);
- zero error of an instrument (for example, a balance that has been improperly set for zero mass with nothing on the weighing pan);
- poor calibration of an instrument (for example, every volt of signal being measured is either too large or too small).

Correcting for systematic errors will improve accuracy; as Jean-Nicolas Nicollet did when re-analyzing the data collected by Pierre Méchain. But sometimes you can only find a systematic error by measuring the same value by a different method.

Let us now look further at accuracy and precision. Precision is a description of random errors; a measure of statistical variability. Accuracy has two definitions: most commonly, it is a description of systematic errors, a measure of statistical bias; but alternatively, the International Organization for Standardization (ISO) defines accuracy as describing both types of observational error mentioned above (preferring the term trueness for the common definition of accuracy). Schematically, this may be represented as:[1]

---

[1] See Wikipedia's entry for accuracy and precision for further details.

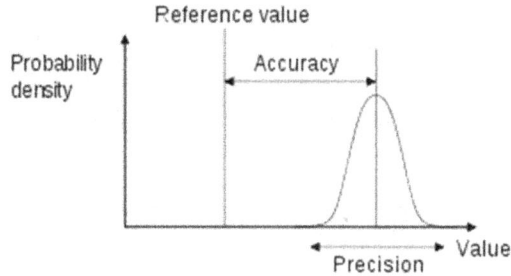

Accuracy is the proximity of a measurement result to the true or reference value; precision is the repeatability or reproducibility of the measurement; that is, the range of measurement results.

In science, the accuracy of a measurement system is the degree of closeness of measurements of a quantity to that quantity's true value. The precision of a measurement system, related to reproducibility and repeatability, is the degree to which repeated measurements under unchanged conditions show the same value. Although the two words precision and accuracy can be synonymous in a colloquial sense, they are deliberately contrasted in the context of measurement science.

A measurement system can be accurate but not precise, precise but not accurate; a measurement system can be neither or both. For example, if an experiment contains a systematic error (a poor calibration), then increasing the sample size generally increases precision but does not improve accuracy. The result would be a consistent, yet inaccurate set of results from the flawed experiment; Méchain's problem in his analysis of the data from the metric survey. Eliminating the systematic error improves accuracy but does not change precision. In addition to accuracy and precision, measurements may also have a measurement resolution, which is the smallest change in the underlying physical quantity that produces a response in the measurement.

A measurement system is considered valid if it is both accurate and precise. Related terms include bias (non-random or directed effects caused by a factor or factors unrelated to the independent variable) and error (random variability). This terminology is also applied to indirect measurements; that is, values obtained by a computational procedure from observed data. In numerical analysis, accuracy is also the nearness of a calculation to the true value; while precision is the resolution of the representation, typically defined by the number of decimal places after the decimal separator.

Mathematical or statistical literature prefers to use the terms bias and variability instead of accuracy and precision, which is the preferred vocabulary of the laboratory scientist: bias is the amount of inaccuracy, and variability is the amount of imprecision. In industrial instrumentation, accuracy is usually the measurement tolerance, or transmission of the instrument and defines the limits of the errors made when the instrument is used in normal operating conditions.

Ideally a measurement device is both accurate and precise, with measurements all close to and tightly clustered around the true value; certainly this should be the case if the experiment has been designed for a particular measurement (as we will see in chapter 8). The accuracy and precision of a measurement process is usually established by repeatedly measuring some traceable reference standard. Such standards are ultimately defined in the International System of Units and maintained by the BIPM (*Bureau internationale des poids et mesures*), near Paris.

This also applies when measurements are repeated and averaged. In this case, the term standard error is applied as: the precision of the average is equal to the known standard deviation of the process divided by the square root of the number of measurements involved in the averaging process. Further, the central limit theorem tells us that the probability distribution of the averaged measurements will be closer to a normal distribution than that of individual measurements.

Precision is sometimes stratified into:

- repeatability: the variation arising when all efforts are made to keep conditions constant by using the same instrument and operator, and repeating measurements during a short time period; and
- reproducibility: the variation arising using the same measurement process among different instruments and operators, and over longer time periods.

A shift in the meaning of accuracy and precision appeared with the publication of the ISO 5725 series of standards in 1994, which is also reflected in the 2008 edition of the International Vocabulary of Metrology (VIM) published by the BIPM; particularly, items 2.13 and 2.14. According to ISO 5725-1, the general term accuracy is used to describe the closeness of a measurement to the true value. When the term is applied to sets of measurements of the same quantity, it involves a component of random error and a component of systematic error. In this case trueness is the closeness of the mean of a set of measurement results to the actual (true) value and precision is the closeness of agreement among a set of results.

ISO 5725-1 and VIM also avoid the use of the term bias, previously specified in BS 5497-1, because it has different connotations outside the fields of science and engineering, as in medicine and law.

Consider the following explanation of measurement uncertainty taken from the chemical literature. The nomenclature is different from that often seen in physics, but this explanation does introduce some useful ideas and figure 3.1 is particularly informative.

Measurement is the process of experimentally obtaining the value of a quantity (that is, a value with an associated unit and quantity, as in x moles per litre). The quantity that is measured is termed the measurand. In chemistry the measurand is usually the content (that is, concentration) of some chemical entity in some solvent. The chemical entity under study is termed the analyte. In principle, the aim of a measurement is to obtain the true value of the measurand. Every effort can be made to optimize the measurement procedure in such a way that the measured value is as close as possible to the true value. However, the measurement results will be just an estimate of the true value, as the actual true value will (almost) always remain unknown. Therefore, we cannot know exactly how near our measured value is to the true value—our estimate always has some associated uncertainty.

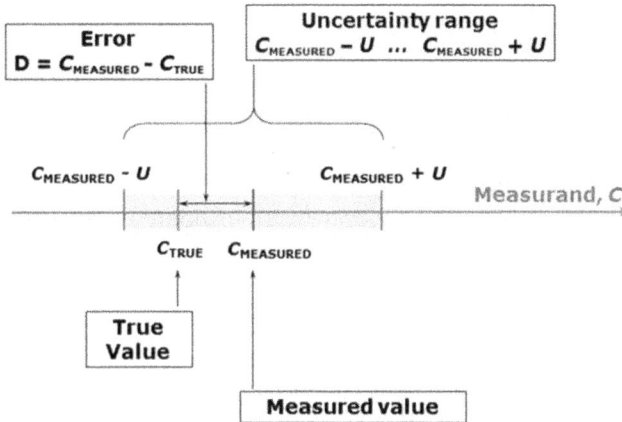

Figure 3.1. Interrelations between the concepts involved in defining the quality of a measurement: the true value (which is hardly known in practice, and certainly not in research), the measured value, the error (the difference between the measured value and the true value; of which the latter can never be known with full certainty, so this quantity is something of an abstraction) and the measurement uncertainty (which may be computed).

The difference between the measured value and the true value is called the error. Errors can be positive or negative. As we saw earlier, error can be regarded as being composed of two parts, random error and systematic error. Like the true value of the measurand, the true value of the error is also not known (hence the need for the probabilistic methodology in the study of measurement uncertainty.

The quality of the measurement result, its accuracy, is characterized by the measurement uncertainty, which defines an interval around the measured value C_{MEASURED}, where the true value C_{TRUE} lies with some estimated probability. The measurement uncertainty U itself is the half-width of that interval and is always non-negative. This is illustrated in figure 3.1, which should be compared with the schematic on page 3-4.

Measurement uncertainty is always associated with some probability, but it is not possible to define an uncertainty interval in such a way that the true value of the measurand lies within it with 100% probability. The measurement uncertainty expressed above, is in some contexts also known as the absolute measurement uncertainty. This means that the measurement uncertainty is expressed in the same units as the measurand. It is sometimes more useful to express measurement uncertainty as relative measurement uncertainty, U_{rel}, which is the ratio of the absolute uncertainty U_{abs} and the measured value y,

$$U_{\mathrm{rel}} = U_{\mathrm{abs}}/y.$$

Relative uncertainty is a dimensionless or unit-less quantity, which is sometimes also expressed as a 'per cent'.

Measurement uncertainty is different from error in that it does not express a difference between two values, and it does not have a sign. Therefore, it cannot be used for correcting the measurement result, and cannot be regarded as an estimate of the error. Instead measurement uncertainty can be thought of as an estimate of what is the highest probable absolute difference between the measured value and the true value. However, both the true value and error (random and systematic) are abstract concepts. Their exact values cannot be determined. But these concepts are nevertheless useful, because their estimates can be determined and hence compared between; for example, different experimenters or different laboratories. As said above, our measured value can only ever be an estimate of the true value.

Further reading

Measurement uncertainty is a vast subject, and if you Google it you will be swamped by responses. You will also discover that there is no single international language or vocabulary of terms and definitions; however, at www.bipm.org/en/publications/guides/vim.html you can find all 108 pages of the international vocabulary of metrology (the VIM), including useful diagrams to explain the terms and concepts. With regard to the points outlined in the last part of this chapter; particularly, with reference to figure 3.1, the ten minute video at www.youtube.com/watch?v=BogGbA0hC3k is well worth watching.

Chapter 4

What is it that we measure, and what does it tell us?

Since we are assured that the all wise Creator has observed the most exact proportions, of number, weight and measure in the make of all things, the most likely way therefore to get any insight into the nature of those parts of the creation, which come within our observation, must in all reason be to number, weight and measure.

Stephen Hales, 1677–1761

Before we look in detail at how measurements are quantified, let us consider briefly some very different types of measurements. In this chapter, we will first consider two standard measurements; that is, where the experimenter is measuring or monitoring one characteristic property of a system while some other property of that system is varied. Then we will look at two non-standard measurements, which although not made in what you might consider to be a scientific laboratory are, in fact, more important (for different reasons) than measurements made in a classic laboratory setting.

4.1 A classic laboratory experiment

In figure 4.1 we see the birefringence induced in a high-pressure sample of carbon dioxide by a large, constant applied electric-field gradient, as a function of the temperature (room temperature to 123 °C) of the gas sample. This data is taken from the author's PhD thesis, and represents several weeks of measurements in the summer of 1979. In chapter 8, we will look in detail at this experiment when we consider the design of an experiment. But here, I wish to draw the reader's attention to the classic nature of this set of measurements; it is what most people envisage as

doi:10.1088/978-1-6817-4433-9ch4   4-1

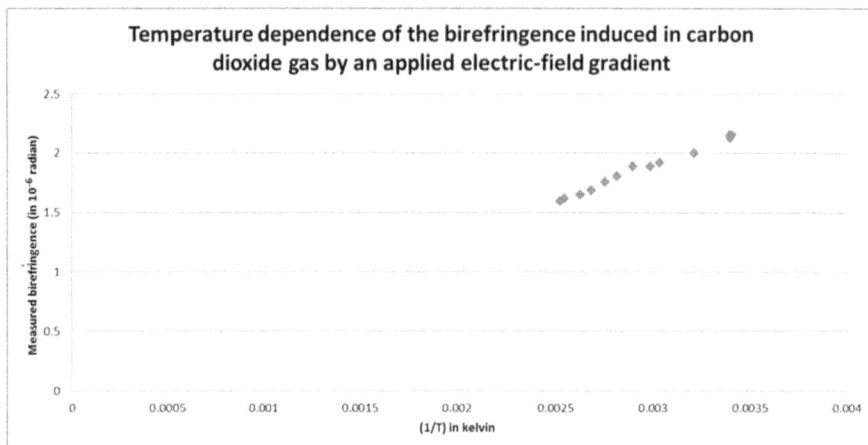

Figure 4.1. A classic laboratory measurement: the temperature dependence of the birefringence induced in a high-pressure sample of carbon dioxide by an applied electric-field gradient. The measured birefringence is plotted against the reciprocal of the absolute temperature to demonstrate the presence of both a temperature dependent term and a temperature independent term in the observed measurement. The data is taken from the author's PhD thesis, and is published in Battaglia M *et al* 1981 *Mol. Phys.* **43** 1015.

an experiment. When the apparatus was working well, a measurement of the induced birefringence took about an hour. However, it took rather longer than this to change the sample temperature as the high-pressure gas cell was made of two concentric cylinders of stainless steel, each 1 m in length, and these tubes were heated by circulating externally heated anti-freeze through the space between the cylinders.

The applied electric-field gradient orients the molecules, and the medium becomes birefringent; so linearly polarized light propagating through the oriented medium becomes elliptically polarized. The induced retardation is, of order, 10^{-6} radian for available laboratory electric-field gradients and pressures of several atmospheres of carbon dioxide. There are two terms that contribute to the measured effect; one of which is temperature-dependent and another which is independent of temperature. Undertaking measurements of the induced birefringence at a variety of temperatures therefore allows one to measure both these contributing factors.

From the data in figure 4.1, one is able to extract both the temperature-dependent term contributing to the measurement; that is, the slope of the line joining the measured points, and the temperature-independent term contributing to the measurement; that is, the intercept on the vertical axis of the line joining the data points. There is noise or scatter on the data displayed in this figure, and this noise will be the major component to determining the uncertainly in the determination of the slope of the line, and will certainly dominate the determination of the intercept as the extrapolation is over such a long distance (the lever arm principle, as can be seen in figure 4.1), and as you go to higher temperatures, the signal decreases and so the signal/noise ratio is not moving in your favour. In actual fact, the uncertainty in the slope of the line gives a quadrupole moment for CO_2 of $-15.3 \pm 0.7 \times 10^{-40}$ cm^2; an uncertainty of 4.5%. Given this level of uncertainty, and the range of the extrapolation, it is not possible to say anything more than that the temperature independent term (determined from the intercept of the data in figure 4.1) contributes less than 4.5% to the measured birefringence at room temperature.

This data is typical of a classic laboratory experiment, where there is some measured parameter, which is a function of temperature, $f(T)$ and is plotted against sample temperature. There are two unknown quantities (the temperature-independent term and the temperature-dependent term), and so we require, at least, two equations to determine both unknowns. The experiment is relatively straightforward and can be repeated to establish reproducibility, and to generate sufficient data to obtain decent statistics on the final derived value; the measurements are made within the known precision of the apparatus as we will discuss in chapter 8.

Let us now consider a straightforward measurement, which has huge significance but cannot be repeated endlessly.

4.2 Precision measurements made infrequently

Figure 4.2 displays data representing the temporal evolution of the masses of the six official copies of the International Prototype of the Kilogram (IPK), with respect to the IPK (which in the data displayed in figure 4.2 is labelled K), and which by definition cannot change its mass. The vertical scale is in micrograms (that is, 10^{-6} g or 10^{-9} kg). For two of the copies or *témoins* of the IPK (N43 and N47), however, there are only two real measurements, as all prototypes were assumed to have the same mass in 1889, and N43 and N47 were assumed to have no deviations from perfection in 1947 when they were first measured. For the remaining data sets (the other mass standards) there are only three data points on the figure as perfection was assumed back in 1889. This assumption of perfection in mass in 1889 and in 1947 is of course nonsense. Each of these mass standards would have had a unique mass at that time—it simply was not measured.

This graph appears simple (six objects, most of which are slowly increasing in mass with time), but this simplicity arises because of a lack of data. In 1889, only the

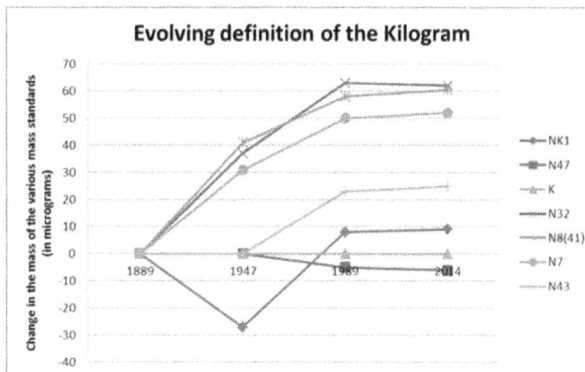

Figure 4.2. The evolution of the masses of the six official copies of the International Prototype of the Kilogram, with respect to the International Prototype of the Kilogram (which in the data displayed here is labelled K). The data is taken from the website of the BIPM (www.bipm.org/en/bipm/mass/ipk/).

IPK would have possessed a mass of precisely 1 kg, because that mass was defined by international law as being precisely 1 kg. By comparison, all the other mass standards would have had a unique, measurable mass, different from 1 kg. Yet for all the sparseness of the data presented in this graph, this image carries a huge responsibility. The measurements are straightforward, one is measuring the force (F) generated by placing a mass (m) in the Earth's gravitational field (g); that is, $F = mg$, Newton's celebrated equation.

The present hierarchy of measurement standards for mass closely resembles a religious dogma. At the highest point is the omnipotent object which is, as far as the world is concerned, indivisible and in which one must believe and have faith, but without ever touching or even seeing it. The perfect mass of this object, precisely 1 kg, is fixed by international diplomatic treaty, the Metre Convention of 1875, and not by mere experiment (and the Metre Convention made no mention of the uncertainty of this definition).

This near-sacred object is the IPK. This memento from the early-days of the Metric System is one of science's most valuable, but also one of its most derided objects; it is kept in a secure vault near Paris surrounded by six identical copies or *témoins*. Thus protected, the IPK reigns supreme over the world's measurements of mass. Every hill of beans, every human, every milligram of medication and recreational drug; in short, the great globe itself and even the smallest of sub-atomic particles, that can be weighed, must be gauged and compared against the mass of that small, glittering, dense object made of an incorruptible alloy, the IPK. And if you believe what you read in the popular press from time to time, this object is mysteriously changing its mass.

These platinum-iridium prototypes are used every so often to monitor how the stability of these mass artefacts is being maintained, and to provide calibration certificates to the rest of the metric world so that we may all see that 'the kilogram' is in its vault and that all is right with the metric world. The good news of the continued stability and perfection of the kilogram then passes down from these closely-guarded objects to their copies in the various Member States of the Metre Convention, and then down the various national pyramids of scientific endeavour and precision measurement to, for example, the humble weighing scale in your bathroom or the electo-mechanical balance next to the vegetables in your local supermarket. In this way, you may be reassured that you are not fooling yourself after a holiday of over-indulgence; as your weighing scale has certificates of calibration linking it to the perfect essence of a kilogram locked away in France. And it is through this unbroken chain of calibration certificates that uncertainty enters mass metrology; the IPK has no uncertainty attached to its mass, but those weighing scales in your local supermarket do have an associated measurement uncertainty.

The IPK will, of course, have a mass that is different from 1 kg, but it cannot be measured. The values of mass along the vertical axis in figure 4.2 give the range of values within which one would be likely to discover the true mass of the IPK, if it could be measured; and will be found, when the kilogram has been redefined by a non-artefact based definition.

So, how much is a kilogram? Well, as it turns out nobody can say for sure; at least, not in a way that won't change ever so slightly over time. And that's not so good for a fundamental standard that the world depends upon to define mass. Of course, such statements in the popular press beg the question, which is never addressed; getting lighter compared to what? The inference is that the kilogram is not a sound unit. That it is getting lighter compared to a more stable mass. But how can this be determined? If there were a more stable mass it would be the IPK, and the present IPK would become just another mass standard. At present, given the artefact-basis of mass metrology, some object has to occupy that solitary position at the apex of the pyramid of mass measurements. The possible instability in the physical characteristics of a metrological artefact is the essential problem with all artefact-based systems of weights and measures, which is why they are re-calibrated every quarter-century or so.

Since it was placed in service in 1889, the IPK has been used during three measurement campaigns or 'periodic verifications of national prototypes of the kilogram'. The most recent such verification was carried out in 2014. Over more than a century, the masses of the official copies are seen to be increasing, with respect to the mass of the IPK (which by definition cannot change its mass and so appears as a horizontal line in figure 4.2 labelled K); mass standard NK1 has lost and gained mass, and standard N47 is losing mass. Interestingly, it is relatively easy to understand how a mass standard can acquire additional mass—we live in polluted cities. But it is not at all straightforward to come up with an explanation of how a platinum–iridium alloy can lose mass with time; although part of that problem disappears if we assume that the mass of mass standard NK1 in 1889 was in fact, well below the mass of the IPK, and similarly for the mass of N47 in 1947.

By definition, the mass of the IPK cannot vary, but given that the masses of its copies are fluctuating, then the mass of the IPK must also be changing, because the IPK and its copies are all stored, and have been stored in the same manner, they are made of the same material and were all made in the same manner. And given that the masses of the copies are mostly increasing, then as they are weighted against the IPK, the mass of the IPK must be decreasing. For the scientists who rely on the continued stability of the base unit of mass of the SI for precise measurements, this inconstant metric mass is a nuisance. Our ability to precisely measure an electric current or a quantity of gas flowing through a pipeline is dependent upon the precision with which the unit of mass is known, and any instability in the precision with which we are able to define the base unit of mass perturbs such calculations; calculations which are worth hundreds of billions of Euros every year.

The problem of the IPK losing mass (it is not) arises because we have an artefact-based system of mass metrology based on seven mass standards that are all becoming increasingly contaminated at different rates. It is a combination of the absorption of contaminants, insufficient data (mass measurements) and the principle of conservation of mass that is creating the confusion.

The SI is the pivot from which hang all measurements, no matter what the area of investigation or the location of the measurement. Whether it is the accuracy of your bathroom scales, the amount of electricity you have consumed in the last month, or the reliability of a petrol pump, there is an unbroken chain of calibration certificates that leads back to realizations of the seven base units of the SI. Science has moved a long way since the IPK became the basis of mass metrology, and this object is becoming increasingly anachronistic; as can be seen in the carefully measured data in figure 4.2. The surest way of stabilizing mass metrology, and to remove our dependence on the drifting values of the masses of a collection or artefacts is to do away with the IPK altogether. Greater precision would also be brought to the SI if the unit of mass were defined by a constant of nature rather than an object; no matter how carefully conserved. The choice of the physics community is to redefine the unit of mass in terms of Planck's constant, and the choice of the chemistry community is to redefine the kilogram via Avogadro's number (for details about the redefinition of the SI, please see Willilams J H 2014 *Defining and Measuring Nature; The make of all things* (San Rafael, CA: Morgan & Claypool).

Let us now consider data that represent many hundreds of measurements made on blood samples from a single patient.

4.3 An overabundance of uncertain data

The data given in figure 4.3 represent routine laboratory measurements made on the blood of someone living with human immunodeficiency virus (HIV). This data set is unique; that is, a set of the same measurement data over the same period from another person living with HIV would look very different. The data in figure 4.3 represent the life of an individual, and no other individual would generate an identical set of results. This is the problem when designing clinical trials to test the efficacy of new medications, which we shall look at in chapter 7. We are all distinct, whereas all carbon dioxide molecules behave in the same way when stimulated in the same manner (see figure 4.1), and all mass standards become coated in pollutants (figure 4.2). But no two humans respond in exactly the same way to the same amount of a single drug; this is termed genetic variability.

The data in figure 4.3(*a*) covers a period of about fifteen years. Each data point represents the measurements made by an automated blood sampler in the laboratory of a large hospital, which takes blood that a nurse has taken from a patient's arm, dilutes it and scatters laser light from the diluted sample; the intensity of scattered light is proportional to the number of CD4 cells in the blood (in one microlitre of blood). The automated apparatus is quoted as having a 10% error, which arises principally in the automated dilution procedure. Each blood sample required a new syringe, and these mass-produced syringes have a 10% uncertainty in their volume (±5%). This uncertainty propagates through to the final value of the number of CD4 cells in one microlitre of the patient's blood.[1]

Obviously, the data in figure 4.3(*a*) is hugely important to the patient involved, and to his or her clinician who is responsible for their successful treatment. Indeed, it is from such data that decisions are taken about whether or not a particular

---

[1] For your interest, the lower the CD4 concentration, the more at risk of illness and death is the patient; and levels below 200 (on the scale in this figure) are considered to be dangerously low.

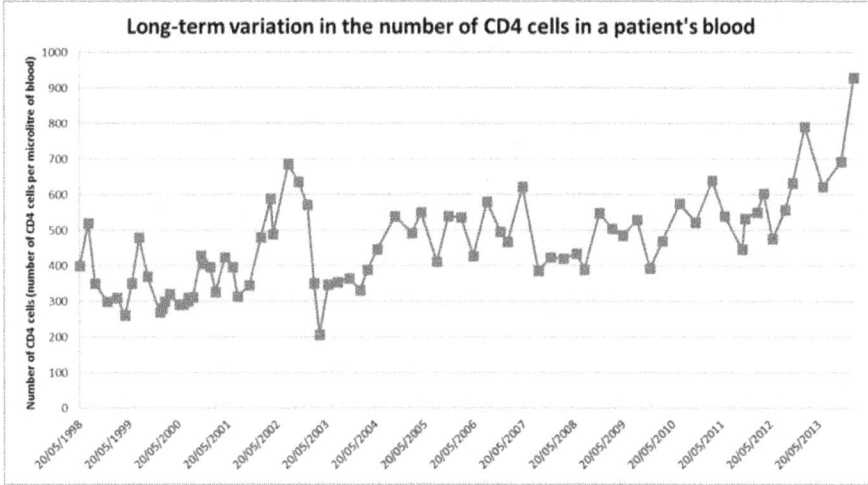

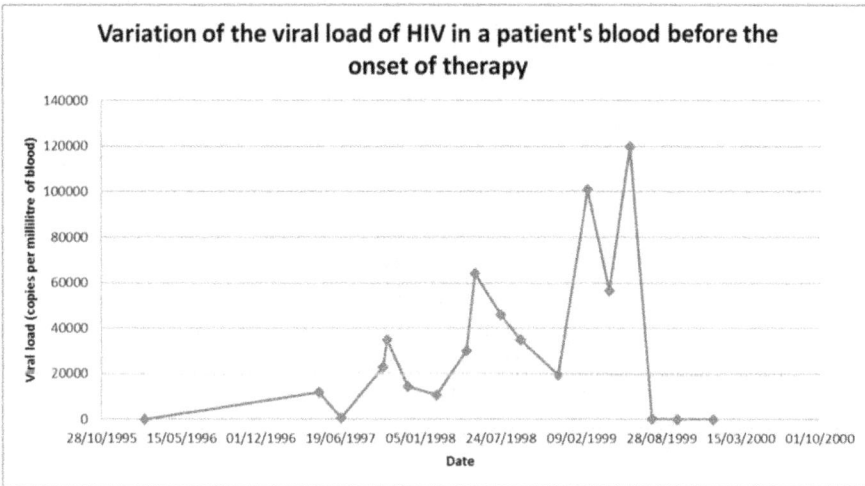

Figure 4.3. (*a*) A more complex set of measurements: the evolution over many years of the number of CD4 cells (an essential component of the human immune system) in a patient living with human immune deficiency virus (HIV). The units are number of CD4 cells per microlitre of blood, and the uncertainty of the measurement is ±5%. (*b*) An even more uncertain (±50%) set of measurements, but which are of great importance to the patient: the evolution of the amount of HIV virus (the viral load) in the patient's blood before the onset of treatment. The units are copies of HIV nucleic acid per millilitre of blood.

treatment or combination of anti-HIV drugs is failing for this patient (the clinician is not trained to consider measurement uncertainties, but he or she can see if the data is trending upwards or downwards with time). But knowing that there is a 10% error on the measured value is very important as there is no point in changing the therapy for a patient who has successive CD4 counts of 510, 460 and 470 (per microlitre of blood), as these values are all within the overlapping error limits. And indeed, we can see from the data in figure 4.3(*a*) how the values of successive measurement can see-saw quite dramatically over six or four month intervals (by as much as ±200 cells per microlitre or up to ±30% of the total number of cells). But we can also see periods in figure 4.3(*a*) when the data was pretty consistent over a period of a year or more, with almost identical values being measured.

And what of the dramatic fluctuations seen in this data set? Why does the data fluctuate so wildly at some times, yet is relatively constant at other times, given that the uncertainty on each measurement is 10%? Well, we know that CD4 levels fluctuate diurnally in all of us; that is, the levels are higher in the morning (perhaps by up to 20%) than in the evening, and that women have higher levels of CD4 cells than men (of the same age) and that pregnant women have very high levels of CD4 cells. Depending upon the sex of the patient who supplied the anonymous data in figure 4.3(*a*), all these factors will operate to generate the fluctuations seen in the data. The diurnal fluctuations will be superimposed on the patient's measurements like a background noise in a classic laboratory measurement. It will always be there as we cannot demand that the patient always give blood at exactly the same time of the day over many years. The other reason the data is fluctuating is that the CD4 cell is the basis of the body's defence against invading micro-organisms and internal problems such as cancer, and the greater the number of infections, the lower the levels of CD4 cells. Figure 4.3(*a*) is also the record of how the patient was feeling over the time period displayed; how they were feeling both physically and psychologically, as stress also lowers CD4 levels.

Figure 4.3(*b*) displays measurements of the viral load (the virus is again HIV) in a patient's blood just before the onset of treatment to control the HIV levels in the patient; this is another routine blood measurement carried out in all large hospitals. A measurement of viral load, is an indirect measurement of the amount of virus in a quantity of blood, and relies on a chemical process called the polymerase chain reaction (PCR). This is something that has revolutionized biology and forensic science in the last 25 years, and won its discoverer, the American biochemist, Kary Mullis (born 1944) a Nobel Prize in 1993.

PCR is useful because the genetic material of every living thing possesses sequences of chemical building blocks that are combined together in a unique manner. PCR exploits the ability of certain chemical catalysts that are present in all living organisms to make exact copies of genetic material. It copies the process which happens when the genetic material is being transferred from one generation of an organism to the next (cell division). Sometimes referred to as 'molecular photocopying', PCR can characterize, analyze, and synthesize any specific piece of genetic material. It works even on complicated mixtures, finding, identifying, and duplicating a particular bit of genetic material from samples of blood, hair, or tissue, or from microbes, animals, or plants, which can be many thousands of years old.

Although the chemical composition of the genetic material of HIV is now well known; even in a patient who has advanced HIV disease, the amount of virus in the blood is too low to be measured directly. So PCR is used to amplify the amount of genetic material to such a degree that it becomes measurable and quantifiable.

To begin the PCR process, an automated blood analyzer will sample a known quantity of a patient's blood, which contains some HIV and hence some HIV genetic material, and mix it with an appropriate label. This label or tag is designed so that it will only combine with the genetic material of HIV. Such a tagged or labelled piece of HIV genetic material is termed the 'template'. It is this template or labelled part of the genetic material of the virus, which is copied. Because the building blocks of the genetic material only combine together following specific rules, they will only combine together in the presence of the template to make copies of the original labelled genetic material. In this way, the original minute quantity of genetic material in the blood can be amplified to make measurable quantities.

There are three basic steps in PCR. First, the genetic material must be removed from the virus. The second step is the labelling process where the specific tag or label is attached to HIV's genetic material. The third stage is chemical amplification. The result of each stage of the amplification process is that the number of original pieces of labelled genetic material doubles. So starting from one piece of genetic material, one can make, 2, 4, 8, 16, 32, ..., pieces at each stage of the chemical amplification. Each cycle of amplification takes only a few minutes, and repeating the process for just a couple of hours can generate millions of copies of a specific genetic material. As in all laboratory measurements, technical limitations apply to PCR. The most important is contamination of a patient's sample with some unwanted genetic material that could also be amplified and generate numerous copies of irrelevant genetic material. The result of such a contaminated amplification will often simply be useless, but sometimes it can lead to wrong conclusions.

Because of the potential for contamination in the PCR process, the uncertainty in the measurement of the viral load is high. The amplification process is at the heart of the PCR, and so the final measurement is given in 'copies' (copies per millilitre of blood). Thus clinicians speak of a viral load of 100 000 copies; see figure 4.3(b). If the viral load is undetectable, this means that there is so little HIV present in your blood that even with amplification, the PCR reaction cannot produce a measurable quantity. Such a result does not mean that there is no HIV present, only that the treatment is working and the amount of virus in the patient's blood is below the precision of the PCR equipment.

The uncertainty on the final measurement can be as high as ±50%. Thus when a viral load is measured at 100 000 copies, the actual value could be as high as 150 000 or as low as 50 000. Thus when two successive measurements are compared, for example, 200 000 and 150 000, is there a difference between these two numbers? A statistician would tell you that it is difficult to see any 'clear daylight' between these two measurements with such overlapping uncertainties. Of course, if one viral load is 200 000 and the next is 20 000, then there is clearly a difference and the anti-HIV therapy is working. This is what is seen in figure 4.3(b), where the viral load measurements increase erratically to the point where treatment is initiated, and then the values fall very quickly.

Not only has the PCR process made it possible to measure the amount of a virus in blood, it is now an essential tool in forensic science—the genetic fingerprint. PCR is routinely used to determine the paternity of a child, and has been used to follow the family relations of ancient Egyptian mummies. Even if the Pharaoh has been dead three thousand years, there is still enough genetic material present in the embalmed body to establish if he or she is related to the mummy in the tomb next door. Yet, as mentioned above, there are significant uncertainties generated in the PCR process, and this uncertainty becomes greater, the poorer the quality of the nucleic acid being amplified. Thus, in forensic science applied to criminal cases, it is arguments about measurement uncertainty that limits the use of PCR technology; are we sufficiently sure or certain (as determined by our laboratory measurements) that we can deprive someone of their liberty (or worse) for many years?

In these clinical measurements, we are a long way from the classic laboratory data seen in figure 4.1 or figure 4.2, yet such unique, personal medical data represent the largest set of measurements made today, and the number of such automated blood-based measurements made each year is increasing by, of order, 10%.

The data given in figure 4.3 may be personal, but such data is used (in an anonymous form) for research. Could one, for example, use such data to explore the 'placebo' effect? The medication is taken daily, so the quantity of active ingredient in the patient's body is constant, so why does the data see-saw in the manner seen in the figure? Could it be the in the summer months, the patient is feeling better than in the winter months and this has an effect upon his or her immune system? The question that has to be asked is how to extract data from such a set of measurements. What does it all mean? Does it mean anything at all ... except of course to the one patient who gave all those blood samples and which form a record of his or her life? One has to be careful, however, about how you look at such complex data. When viewed in one way, it may tell you something, but when viewed in another sense, the meaning might change completely (see figure 4.4).

Now for something completely different; measurements and data sets that are studied in detail by those who run our society.

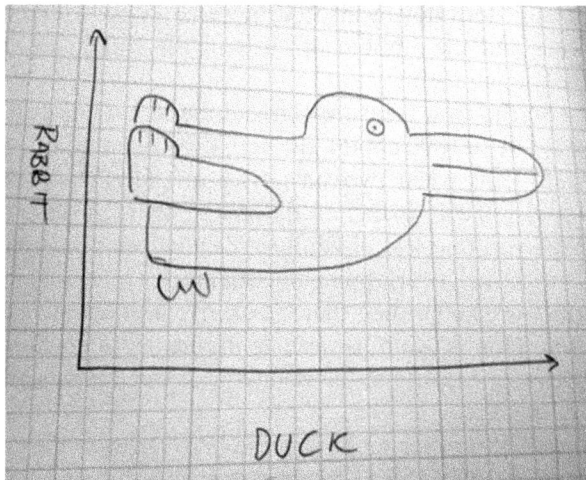

Figure 4.4. An image to remind us that a single set of data points can have a number of interpretations, depending upon how one looks at it.

4.4 What makes the world go around?

If you have a personal pension, you have an interest in knowing something about how the Stock Market functions. All pension funds are in some way linked to the Stock Market; either to 'blue chip' shares that usually fluctuate in value slowly and by small amplitudes, or to riskier stocks whose value can be all over the place.

Since the 1980s, the Stock Market has become, for the vast majority, the only method of growing an investment, apart from investing in property. The data in figure 4.5 is derived from the average value of a company on a daily basis, for the hundred largest companies in the UK (the FTSE 100). Figure 4.5(a) shows how this

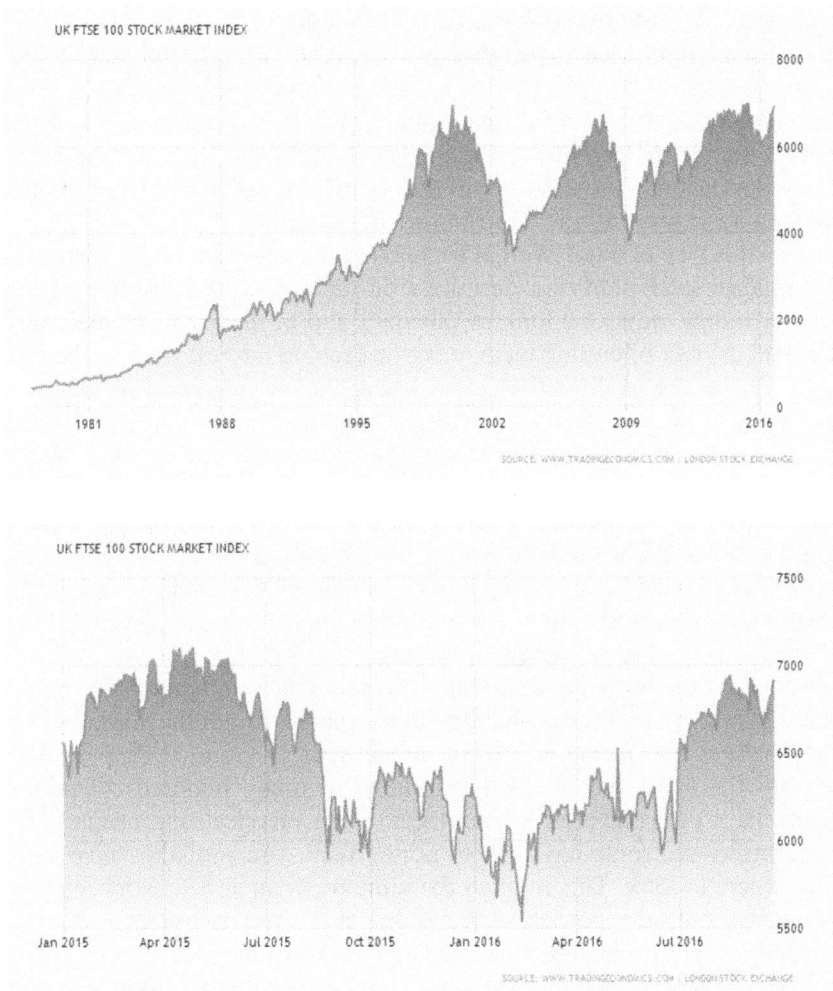

Figure 4.5. A very complex set of data (or observations, but they can still be considered as measurements): the evolution of the FSTE100 share average: (a) over the last 35 years, and (b) January 2015–August 2016. What is contributing to these fluctuations? What on earth would be the uncertainty on these observations, and how would you estimate it? Data taken from www.tradingeconomics.com.

share average has evolved to August 2016, and figure 4.5(*b*) shows a more detailed section of part (a), covering the period January 2015 to the end of August 2016.

Looking at the data set in figure 4.5(*a*), which are of huge significance and importance to the national and the international economy and which affect the lives of all of us, one can clearly see periods of spectacular growth, but also periods of spectacular loss. Up until 2001, there was a near exponential growth in the value of the FTSE100 (and the wealth of the individuals who owned those hundred companies. Then after September 2001, the market crashed taking a year to reach its nadir. It then recovered over about seven years to its 2001 value before crashing as a consequence of the sub-prime banking crisis in the USA. It then rose in a complex set of curves to the value it has today.

Interestingly, the rate of recovery; that is, the rate of growth of the FTSE 100, once underway, can be seen to be relatively constant, varying from about 400 units per year to about 600 units per year. What these rates of increase mean is, however, much more difficult to say. The great engine of the world economy is dependent upon a vast number of complex things; from an investor's optimism to a criminal trader's insider dealing, but all we see in the complex set of curves given in figure 4.5 is the overall average of all the contributing parameters.

Who uses this type of data? Well, it is the basis of capitalism and the world is now wholly capitalist; so all politicians scrutinise this data. Our Prime Minister (or rather his or her economic advisors) look at this data and at the levels of unemployment and calculate the risk of cutting income tax or keeping interest rates at their present level, and what is the likelihood of them losing the coming general election. On a personal level, if your pension plan is tied to the FSTE 100 and it falls, as it did earlier this year, from 6500 to 5700 then your pension pot is worth 5700/6500 of what it was before the markets fell. This is why many inspect the daily rises and falls of the Stock Market, waiting for it to stabilize so as to cash in their pension pots.

The ups and downs of the data seen in figure 4.5 are difficult to interpret fully, except of course for the major collapses seen in figure 4.5(*a*), but we and our futures are in many ways dependent upon these complex curves. We are all caught up in the coils of the data graphed in these two figures.

What interests the investors who study the data given in figure 4.5 is actually an extrapolation or a sense of the probability as to whether or not the market will rise or fall. They believe that signal is present in the data, but buried deep in the noise arising from the fears, worries and ambitions of many hundreds of millions of individuals. How does one even go about seeking for this electronic needle of interest in the enormous electronic haystack of noise? Well, you probably have to have a good idea where to look. This is much the same problem as a researcher looking to observe a new sub-atomic particle, or the problem our secret services have in seeking to identify the email exchanges and telephone conversations of terrorists planning murder and mayhem in the cacophony and complexity of daily Internet traffic. It is not easy, but progress is being made.

As mentioned earlier, measurements are made in myriad forms for myriad reasons. Some measurements are what one might consider to be classic scientific experiments; a unique piece of fundamental research where the result is of interest to a handful of scientists. And some are made routinely for medical reasons, where there is no fundamental law of physics to be decided or measured, but it can be a matter of life and death for an individual whose body is the subject of the measurement. Then there are the measurements and data associated with society, how the world works. Measurements of complex parameters, which have the values that are observed for enormously complex, inter-connected reasons that economists are not too good at divining, but which affect and effect us all. Figure 4.5 represents the complex interaction of multi-dimensional parameters, which with time keep providing new paradoxes, or more precisely, new data since we cannot know the 'right answer' to any question or the 'right value' of any phenomenon.

Chapter 5

Measurement uncertainty

Errors using inadequate data are much better than those using no data at all.
Charles Babbage, 1792–1871

5.1 Uncertainty

All experimentation is (or should be) a voyage of discovery, or at least an investigation of something about which the investigator is uncertain and wishes to be more certain. The subject of uncertainty is vast. It involves imperfect and/or unknown information. In other words, it is a term used in subtly different ways in a number of very different fields, from insurance and philosophy to physics and information science. It applies to predictions of future events, to physical measurements that have already been made, and to the unknown.

Even among specialists in the quantitative fields, the definitions of uncertainty and risk and their measurement are varied, and this variety becomes greater when we consider these terms as used by non-specialists:

- Uncertainty: The lack of certainty. A state of having limited knowledge where it is impossible to exactly describe the existing state, a future outcome, or more than one possible outcome.
- Measurement of uncertainty: A set of possible states or outcomes where probabilities are assigned to each possible state or outcome; this includes the application of a probability density function to continuous variables.
- Risk: A state of uncertainty where some possible outcomes have an undesired effect or significant loss.
- Measurement of risk: A set of measured uncertainties where some possible outcomes are losses and the magnitudes of those losses.

doi:10.1088/978-1-6817-4433-9ch5

Of course, these definitions create as many questions as they answer. Perhaps the best recent example of the confusion arising from the term uncertainty is due to Donald Rumsfeld (United States Secretary of Defense, 2001–06) when he was commenting in a news briefing (12 February 2002) about the lack of evidence linking the government of Iraq with the supply of weapons of mass destruction to terrorist groups.

Secretary Rumsfeld stated:

Reports that say that something hasn't happened are always interesting to me, because as we know, there are known knowns; there are things we know we know. We also know there are known unknowns; that is to say we know there are some things we do not know. But there are also unknown unknowns—the ones we don't know we don't know. And if one looks throughout the history of our country and other free countries, it is the latter category that tends to be the difficult one.

Many thoughtful people are still trying to comprehend what was behind this gnomic statement. But then, you can never be certain about uncertainty.

Quantitative uses of the terms uncertainty and risk are fairly consistent through fields such as probability theory, actuarial science, and information theory. Vagueness and ambiguity are sometimes described as second-order uncertainty, where there is uncertainty even about the definitions of uncertain states or outcomes; as happens when we look at outcomes in the social sciences. The difference here is that this uncertainty is about subjective definitions and concepts, rather than an objective fact of nature.

The state of uncertainty could also be purely a consequence of a lack of knowledge, but of facts that could be observed. That is, there may be uncertainty about whether a new experimental apparatus will work, but this uncertainty can be removed with further analysis and experimentation. At the sub-atomic level, however, uncertainty may be a fundamental and unavoidable property of the Universe. In quantum mechanics, the Heisenberg Uncertainty Principle puts limits on how much an observer can ever know about the position and velocity of a quantum particle. This is the way nature is, at this level.

Uncertainty in science is something guaranteed to generate controversy, and has led to science being ridiculed in the popular press. This is due, in part, to the diversity of the public audience, and the tendency for scientists to communicate ideas poorly and ineffectively. In addition, there are often many scientific voices attempting, with varying degrees of success to provide meaningful input on a single topic; particularly, on emotive topics such as climate change. In extreme cases, depending on how an issue is reported in the public domain, discrepancies between outcomes of multiple scientific studies due to methodological differences, and their subsequent statistical analysis, are sometimes interpreted by the public as representing a lack of consensus in a situation where a consensus does in fact exist.

Indeterminacy of measurement can be said to apply to situations in which not all the parameters of the system and their interactions are fully characterized (the validity of a measurement), whereas ignorance refers to situations in which it is not known what there is to know. Such unknowns, indeterminacies and lack of information that exist in science are often transformed by the popular press into uncertainty when reported to the public in order to make the science more manageable or to tell a better 'story', since scientific indeterminacy and ignorance are difficult enough concepts for scientists to convey even among themselves. So uncertainty is often interpreted by the public as ignorance. This is why, when scientists are seeking to communicate subjects that are poorly understood, yet are of considerable interest to the public, for example, mobile phone radiation and public health, they should take particular care when specifying uncertainties associated with them, and the precision of measurements.

5.2 Uncertainty in measurements

As we saw earlier, measurement uncertainty is a non-negative parameter that describes the dispersion of the values attributed to a measured quantity. All measurements are subject to uncertainty, and a measurement result is complete only when it is accompanied by a statement of the associated uncertainty; this uncertainty has a probabilistic basis and reflects incomplete knowledge of the quantity value.

As mentioned previously, the measurement uncertainty is often taken as the standard deviation of a known probability distribution over the possible values that could be attributed to a measured quantity. Relative uncertainty is the measurement uncertainty relative to the magnitude of a particular single choice for the value for the measured quantity, when this choice is non-zero. This particular single choice is usually called the measured value, which may be optimal in some well-defined sense; for example, a mean, median, or mode. Thus, the relative measurement uncertainty is the measurement uncertainty divided by the absolute value of the measured value, when the measured value is not zero.

The purpose of measurement is to provide information about a quantity of interest, which is called the measurand. For example, the measurand might be the value of the molecular quadrupole moment of a molecule, the force generated by a new type of jet engine, or the level of 'bad' cholesterol in your blood. No measurement is exact, and no measurement is free from uncertainty. When a quantity is measured, the outcome depends on the measuring system, the measurement procedure, the skill of the operator, the environment, and a host of other factors. Even if the quantity were to be measured several times, in the same way and in the same circumstances, a different measured value would in general be obtained each time, assuming the measuring system has sufficient resolution or precision to distinguish between the values.

The dispersion of the measured values would relate to how well the measurement is performed. Their average would provide an estimate of the true value of the quantity that generally would be more reliable than a single (albeit well done) measured value. The dispersion and the number of measured values would provide information relating to the average value as an estimate of the true value. However, the true value will almost always be unknown.

The measuring system may provide measured values that are not spread about the true value, but about some value offset from it. If the instrument measuring the data has been miscalibrated, it will work well, but the results derived from the device will all be offset from the true value by the faulty calibration. This was one of the problems faced by one of the two surveyors who derived the value of the metre in the 1790s; the Borda Circle used by Pierre Méchain was worn, and this gave a systematic distortion of his measurements.

Measurement uncertainty has important economic consequences for calibration and measurement activities. In calibration reports generated by national metrological laboratories, the magnitude of the uncertainty associated with the calibration is often taken as an indication of the quality of the laboratory making the calibrations; with smaller values of uncertainty being indicative of higher standing and of higher cost. The *Guide to the Expression of Uncertainty in Measurement* (GUM) is the definitive document on this subject. The GUM has been adopted by organizations responsible for the international standardization of science and technology, and we will look at it in detail in the next chapter.

The above discussion concerns the direct measurement of a quantity. A weighing scale converts a measured extension of a spring into an estimate of the measurand; for example, the weight of a person on the scale. The particular relationship between extension and weight is determined by the calibration of the weighing scale. A measurement model then converts such a quantity value into the corresponding value of the measurand.

There are many types of measurement, and therefore many measurement models. A simple measurement model (for example, for a weighing scale, where the mass is proportional to the extension of the spring) might be sufficient for everyday use. Alternatively, a more sophisticated model of a weighing device, involving additional effects such as air buoyancy (the object being weighed in air will have displaced a volume of air and so by Archimedes' Principle, its weight will change) is capable of providing more appropriate results for industrial or scientific purposes; for example, greater precision as in the measurement of the masses of the mass standards seen in figure 4.2. In general there are often several different quantities, temperature, humidity and displacement that contribute to the definition of the measurand, and these need to be measured to define the measurand.

As well as unanalyzed (raw) data representing measured values, there is another form of data that is frequently needed in a measurement model. Such data relate to quantities representing physical constants, such as the speed of light, c, or the charge on the electron, e, each of which is also known imperfectly, but with greater precision. The items required by a measurement model to define a measurand are known as input quantities in a measurement model. The model is often referred to as a functional relationship, and the output quantity in a measurement model is the measurand of interest.

In the physical sciences, the uncertainty of a measurement, when explicitly stated, is given by a range of values likely to enclose the true value. This may be denoted graphically by error bars, or by the following notations: measured value ± the uncertainty.

Parentheses are the usual means of presenting the ± notation; for example, in stating a length one rarely encounters a fraction ($5\frac{1}{2}$metre), it would be written 5.5 m or 5.50 m, which by convention would mean that the length being investigated was precise to one tenth of a metre or one hundredth of a metre, respectively. This precision is symmetrically distributed about the last written digit; so, we could also write 5.5 ± 0.05 and 5.50 ± 0.005 or 5.50(5) and 5.500(5). If the precision is within two tenths, the uncertainty is ± one tenth, and needs to be stated explicitly. The numbers in parenthesis apply to the numeral left of themselves, and are not part of that number, but part of a notation of uncertainty, and apply to the least significant digit or figure. For instance, 1.007 94(7) represents 1.007 94 ± 0.000 07, while 1.007 94(72) represents 1.007 94 ± 0.000 72. This notation is the basis of scientific reporting.

A reading of 8000 m, with trailing zeroes and no decimal point, is ambiguous; the trailing zeroes may or may not be intended as significant figures. To avoid this ambiguity, the number could be represented as 8.0×10^3 m indicating that the first zero is significant, while 8.000×10^3 m indicates that all three zeroes are significant. However, these rules are not always followed in reporting results. Hence the need for international standardization in nomenclature and uncertainty evaluations; if you are being treated for some chronic illness, requiring regular blood tests, will an emergency blood test you have to make while on holiday be meaningful?

Often, the uncertainty of a measurement is found by repeating the measurement often enough to generate an estimate of the standard deviation of the values. Then, any single value has an uncertainty equal to the standard deviation of the ensemble. However, if the values are averaged, then the mean measurement value has a much smaller uncertainty, equal to the standard error of the mean, which is the standard deviation divided by the square root of the number of measurements. However, such a simplistic statistical analysis of measured results neglects systematic errors.

In probability theory, the normal or Gaussian distribution (as seen in figure 2.2) is a continuous probability distribution used to describe the results of measurements (from the physical size of students in a classroom to the distribution of galaxies in some part of the Universe). Normal distributions are important in statistics, and are often used to represent real-valued random variables whose distributions are not known. The normal distribution is useful because of the central limit theorem, proved by Pierre Simon de Laplace, which in its most general form, including finite variance, states that averages of random variables independently drawn from independent distributions converge to the normal distribution; that is, become normally distributed when the number of random variables is sufficiently large. Physical quantities that are expected to be the sum of many independent processes (such as measurement errors) often have distributions that are nearly normal. Moreover, many results and methods (such as propagation of uncertainty and least-squares parameter fitting) can be derived analytically in explicit form, when the relevant variables are normally distributed.

The normal distribution is sometimes termed a bell-shaped curve (see figure 2.2). However, many other distributions are also bell-shaped (such as the Cauchy, Student's *t*, and logistic distributions).

The probability density of the normal distribution is:

$$f(x|\mu, \sigma^2) = \left(1/\sqrt{2\sigma^2\pi}\right)\exp -(x-\mu)^2/2\sigma^2,$$

where μ is mean or expectation of the distribution (and also its median and mode), σ is standard deviation, and σ^2 is variance.

From the properties of such probability distributions, we are able to make statements about measurement uncertainty when the uncertainty represents the standard error of the measurement; for example, from figure 2.2 when the true value of the measured quantity falls within the stated uncertainty range 68.3% of the time. Such a probability distribution has the advantage that the quoted standard errors are easily converted to 68.3% (one sigma or one standard deviation distributed about the mean), 95.4% (two sigma or twice the standard deviation distributed about the mean), or 99.7% (three sigma or three times the standard deviation distributed about the mean) confidence intervals.

In this context, uncertainty depends on both the accuracy and precision of the measuring instrument. The lower the accuracy and precision of the instrument, the larger the uncertainty of the measurement.

Formally, the output quantity, denoted by Y, about which information is required, is often related to input quantities, denoted by $X_1, ..., X_N$, about which information is available, via a measurement model in the form of,

$$Y = f(X_1, ..., X_N)$$

where f is known as the measurement function. A general expression for a measurement model is,

$$h(Y, X_1, ..., X_N) = 0.$$

It is taken that a procedure exists for calculating Y given $X_1, ..., X_N$, and that Y is uniquely defined by this equation.

The true values of the input quantities $X_1, ..., X_N$, are unknown. In the GUM approach to measurement uncertainties $X_1, ..., X_N$, are characterized by probability distributions and treated mathematically as random variables. These distributions describe the respective probabilities of their true values lying in different intervals, to which they are assigned based on available knowledge concerning $X_1, ..., X_N$. Sometimes, some or all of $X_1, ..., X_N$ are interrelated and the relevant distributions apply to these quantities taken together.

Consider estimates $x_1, ..., x_N$, respectively, of the input quantities $X_1, ..., X_N$, obtained from a variety of sources. The probability distributions characterizing $X_1, ..., X_N$ are chosen such that the estimates $x_1, ..., x_N$, respectively, are the expectations of $X_1, ..., X_N$. Moreover, for the ith input quantity, consider a so-called standard uncertainty, given the symbol $u(x_i)$, defined as the standard deviation of the input quantity X_i. This standard uncertainty is said to be associated with the (corresponding) estimate x_i.

The use of available knowledge to establish a probability distribution to characterize each quantity of interest applies to the X_i and also to Y. In the latter case, the characterizing probability distribution for Y is determined by the measurement model together with the probability distributions for the X_i. The determination of the probability distribution for Y from this information is known as the propagation of distributions. Figure 5.1 depicts a measurement model $Y = X_1 + X_2$, in the case

Figure 5.1. Figure depicting a measurement model, $Y = X_1 + X_2$ in the case where both inputs X_1 and X_2 are each characterized by a (different) rectangular, or uniform, probability distribution. In this case, the output function has a symmetric trapezoidal probability distribution.

where X_1 and X_2 are each characterized by a (different) rectangular, or uniform, probability distribution Y has a symmetric trapezoidal probability distribution (figure 5.1).

Once the input quantities $X_1, ..., X_N$ have been characterized by appropriate probability distributions, and the measurement model has been developed, the probability distribution for the measurand Y has been fully specified. In particular, the expectation of Y is used as the estimate of Y, and the standard deviation of Y as the standard uncertainty associated with this estimate.

Often an interval containing Y with a specified probability is required. Such an interval, a coverage interval, can be deduced from the probability distribution for Y. The specified probability is known as the coverage probability. For a given coverage probability, there is more than one coverage interval. The symmetric coverage interval is an interval for which the probabilities of a value to the left and the right of the interval are equal. The shortest coverage interval is an interval for which the length is shortest over all coverage intervals having the same coverage probability.

Prior knowledge about the true value of the output quantity Y can also be incorporated into the model; for example, a bathroom scale, the fact that the person's mass is positive, and that it is the mass of a person, rather than that of a motor car that is being measured. Both these conditions constitute prior knowledge about the possible values of the measurand. Such additional information can be used to provide a probability distribution for Y that can give a smaller standard deviation for Y and hence a smaller standard uncertainty associated with the estimate of Y.

The uncertainty of the result of a measurement generally consists of several components. These components are regarded as random variables and may be grouped into two categories according to the method used to estimate their numerical values.

5.3 Type A and Type B uncertainty

Knowledge about an input quantity X_i may be inferred from repeated measured values (this is termed a Type A evaluation of uncertainty: where components of the uncertainty are evaluated by statistical methods), or scientific judgement or other information concerning the possible values of the quantity (this is termed a Type B evaluation of uncertainty: where components of the uncertainty are evaluated by other means; for example, by analyzing a probability distribution).

In Type A evaluations of measurement uncertainty, the assumption is made that the distribution best describing an input quantity X, given repeated independently measured values of this quantity, is a Gaussian distribution. X then has an expectation or expected value (see below) equal to the average measured value and a standard deviation equal to the standard deviation of the average. When the uncertainty is evaluated from a small number of measured values, the corresponding distribution can be taken as a t-distribution.

For a Type B evaluation of uncertainty, it is often the case that the only available information is that X lies in a specified interval $[a,b]$. In such a case, knowledge of the quantity can be characterized by a rectangular probability distribution with limits a and b. If additional information were available, a probability distribution consistent with that additional information would be used.

Sensitivity coefficients $c_1, ..., c_N$ describe how the estimate y of Y would be influenced by small changes in the estimates $x_1, ..., x_N$ of the input quantities $X_1, ..., X_N$. For the measurement model $Y = f(X_1, ..., X_N)$, the sensitivity coefficient c_i equals the partial derivative (first order) of f with respect to X_i evaluated at $X_1 = x_1$, $X_2 = x_2$, etc. For a linear measurement model such as,

$$Y = c_1 X_1 + \cdots + c_N X_N,$$

with $X_1, ..., X_N$ independent, a change in x_i equal to $u(x_i)$ would give a change $c_i u(x_i)$ in y. This statement would generally be approximate for measurement models $Y = f(X_1, ..., X_N)$. The relative magnitudes of the terms $|c_i| u(x_i)$ are useful in assessing the respective contributions from the input quantities to the standard uncertainty $u(y)$ associated with y.

The standard uncertainty $u(y)$ associated with the estimate y of the output quantity Y is not given by the sum of the $|c_i| u(x_i)$, but by these terms combined in quadrature, namely by

$$u^2(y) = c_1^2 u^2(x_1) + \cdots + c_N^2 u^2(x_N),$$

an expression that is generally approximate for measurement models $Y = f(X_1, ..., X_N)$. This is termed the law of propagation of uncertainty. When the input quantities X_i contain dependences, the above formula is augmented by terms containing covariances, which may increase or decrease $u(y)$.

By propagating the values of the variance (the squared deviation of a random variable from its mean), of the various components of the total uncertainty through a function relating the components to the measurement result, the combined measurement uncertainty is given as the square root of the resulting variance. The simplest form is the standard deviation of a repeated measurement; that is, a measurement of one unknown by an apparatus where all other parameters that influence the measured unknown are fully defined.

5.4 Propagation of uncertainty

To apply probabilistic methods to estimating errors and uncertainty in laboratory experiments, one must be able to look at the contributions of each and every element of the experiment to the overall uncertainty in the final result. This is the concept of the propagation of uncertainty.

Most commonly, the uncertainty on a quantity is expressed in terms of the standard deviation, σ, and the positive square root of variance, σ^2. The value of a quantity and its error are then expressed as an interval $x \pm u$. If the statistical probability distribution of the variable is known, or can be assumed, it is possible to derive confidence limits to describe the region within which the true value of the variable may be found. For example, the 68.3% confidence limits for a one-dimensional variable belonging to a normal distribution are $\pm$ one standard deviation from the true value; that is, there is approximately a 68.3% probability that the true value lies in the region $x \pm \sigma$.

Consider an experiment; we are studying a simple electric circuit, and knowing the electrical resistance (R) of the circuit and its measurement uncertainty, and the current flowing through the circuit (I) together with its measurement uncertainty, we wish to measure the voltage (V) using Ohm's law ($V = IR$). Knowing the uncertainties (standard deviations) in I and R, what is the uncertainty in V? In other words, given a function relationship between several measured variables; that is, $Q = f(x, y, z)$, what is the uncertainty in Q if the uncertainties in x, y and z are known?

To answer this question, we must assume that when we talk about the uncertainties in a measured variable that the value of that measured variable represents the mean of a Gaussian or normal distribution, and that the uncertainty in this variable is the standard deviation (σ) of that Gaussian or normal distribution.

To calculate the variance in Q as a function of the variances in; for example, x and y, we use the following (for further details see, for example, the entry for propagation of errors in the online encyclopaedia Wikipedia):

$$Q_Q^2 = \sigma_x^2 \left(\frac{\partial Q}{\partial x} \right)^2 + \sigma_y^2 \left(\frac{\partial Q}{\partial y} \right)^2 + 2\sigma_{xy} \left(\frac{\partial Q}{\partial x} \right)\left(\frac{\partial Q}{\partial y} \right).$$

If the variables, x and y are uncorrelated, that is, $\sigma_{xy} = 0$ then the last term in the above equation vanishes. However, if the variables are correlated, as they would be in the above example of using Ohm's law to determine the voltage in a circuit knowing the resistance and the current, then the last term in the above equation does not vanish and it must be evaluated.

In chapter 8, we will encounter an experiment to measure the molecular quadrupole moment of gaseous molecules. One of the major sources of uncertainty with the measurement is the total amount of light arriving at the detector (one is searching for a signal oscillating at a frequency ω against a large noisy (but un-modulated) background. The total light level at the detector arises from imperfections in the polarizing and the analyzing polarizers and in strain in the windows of optical components that leads to depolarization of the initially linearly-polarized light beam. In this experiment, the uncertainty in the measurement of the signal at frequency ω and the static background would be uncorrelated; whereas the uncertainty in the measurement of the signal at ω and in the uncertainty of the measurement of the pressure of gas in the measurement cell would be correlated.

The average or mean of several measurements, each with the same uncertainty, σ, is given by:

$$\mu = (x_1 + x_2 + \cdots x_n)/n,$$

and the uncertainties may be written as:

$$\sigma_\mu^2 = \sigma_{x1}^2\left(\frac{\partial\mu}{\partial x1}\right)^2 + \sigma_{x2}^2\left(\frac{\partial\mu}{\partial x2}\right)^2 + \cdots \sigma_{xn}^2\left(\frac{\partial\mu}{\partial xn}\right)^2 = \sigma^2\left(\frac{1}{n}\right)^2 + \sigma^2\left(\frac{1}{n}\right)^2 + \cdots \sigma^2\left(\frac{1}{n}\right)^2,$$

which may be written as

$$\sigma_\mu = \sigma/\sqrt{n}.$$

This is the well-known formula for the error in the mean. It tells us that the precision in the final measurement only increases with the square root of the number of measurements. The standard deviation, σ, is related to the probability density function or pdf (usually a normal distribution) from which the values are taken, and σ does not get smaller as we combine measurements.

5.5 Uncertainty evaluation

The main steps involved in the estimation of measurement uncertainty constitute formulation and calculation; the latter stage consisting of error propagation. The formulation stage constitutes:
1. defining the output quantity Y (the measurand),
2. identifying the input quantities upon which Y depends,
3. developing a measurement model relating Y to the input quantities, and
4. on the basis of available knowledge, assigning probability distributions, Gaussian, rectangular, etc, to the input quantities (or a joint probability distribution to those input quantities that are not independent).

The calculation stage consists of propagating the probability distributions for the input quantities through the measurement model to obtain the probability distribution for the output quantity Y, and then using this distribution to obtain:
1. the expectation of Y, taken as an estimate y of Y,
2. the standard deviation of Y, taken as the standard uncertainty $u(y)$ associated with y, and
3. a coverage interval containing Y with a specified coverage probability.

The propagation stage of uncertainty evaluation is also known as the propagation of distributions, various approaches for which are available, including:
- the GUM uncertainty framework, constituting the application of the law of propagation of uncertainty, and the characterization of the output quantity Y by a Gaussian or a t-distribution,
- analytic methods, in which mathematical analysis is used to derive an algebraic form for the probability distribution for Y, and

- a Monte Carlo method, in which an approximation to the distribution function for Y is established numerically by making random draws from the probability distributions for the input quantities, and evaluating the model at the resulting values.

For any particular problem in uncertainty evaluation, any one of these three approaches may be used, however, the first approach is generally approximate, while the second approach is exact, and the third approach provides a solution with a numerical accuracy that can be controlled.

5.6 Probability

As we saw earlier, the great contribution of Gauss to the concept of measurement uncertainty was to introduce the idea of a distribution of possible outcomes of a measurement process; that is, the observed or measured value would be distributed about the true value. In this, Gauss was extending the ideas of the French mathematician, and one of the creators of the Metric System, Pierre Simon de Laplace, who was the first to combine Newtonian mechanics and probability theory (in his *Théorie Analytique des Probabilités* of 1812).

Probability is the branch of mathematics concerned with the analysis of random variables, stochastic processes, and random phenomena; that is, modelling of non-deterministic events or measured quantities that may either be single occurrences or evolve over time in an apparently random fashion. It is not possible to predict precisely the results of random events, however, if a sequence of individual events, such as coin flipping or the rolling of dice, is influenced by other factors, such as friction, it will exhibit certain patterns, which can be studied and predicted. Two representative mathematical results describing such patterns are the law of large numbers and the central limit theorem.

As a mathematical foundation for statistics, probability theory is essential to many activities that involve quantitative analysis of large sets of data. Methods of probability theory also apply to descriptions of complex systems given only partial knowledge of their state, as in statistical mechanics. The great discovery of 20th Century physics was the probabilistic nature of physical phenomena at atomic scales, described in quantum mechanics.

The mathematical theory of probability has its roots in the world of gaming and gambling; attempts to analyze games of chance by Gerolamo Cardano in the 16th Century, and by Pierre de Fermat and Blaise Pascal in the 17th Century (the 'problem of points'). In 1657, Christiaan Huygens published a book on the modelling of games of chance (*De ratiociniis in ludo aleae*), and in 1812 Pierre Simon de Laplace published the first definitive text on probability. Initially, probability theory considered discrete events, and its methods were mainly combinatorial. Eventually, analytical considerations required the incorporation of continuous variables into the theory.

Consider an experiment that can produce a number of outcomes. The set of all outcomes is called the sample space of the experiment. The power set of the sample space is formed by considering all possible results. For example, rolling an honest die produces one of six possible results. One collection of possible results corresponds to obtaining an odd number. Thus, the subset (1,3,5) is an element of the sample space of die rolls. These collections are called events. In this case, (1,3,5) is the event that the die falls on an odd number. If the results that actually occur fall in a given event, that event is said to have occurred.

Probability is a way of assigning an event a value between zero and one, with the requirement that the event be made up of all possible results; in our example, the event (1,2,3,4,5,6) be assigned a value of one. To qualify as a probability distribution, the assignment of values must satisfy the requirement that if you look at a collection of mutually exclusive events (events that contain no common results; for example, the events (1,6), (3), and (2,4) would be mutually exclusive for our present purpose), the probability that one of the events will occur is given by the sum of the probabilities of the individual events.

The probability that any one of the events (1,6), (3), or (2,4) will occur is 5/6. This is the same as saying that the probability of event (1,2,3,4,6) is 5/6. This event encompasses the possibility of any number except five being rolled. The mutually exclusive event (5) has a probability of 1/6, and the event (1,2,3,4,5,6) has a probability of 1; that is, absolute certainty.

Discrete probability theory deals with events that occur in countable sample spaces; for example, throwing dice, experiments with decks of cards, random walk, the tossing of coins, and studies of radioactive decay. Initially, the probability of an event occurring was defined as the number of cases favourable for that event divided by the number of total possible outcomes in an equally probable sample space. For example, if the event of interest is the occurrence of an even number when a die is rolled, the probability is given by 3/6, since three faces out of the six have even numbers and each face has the same probability of appearing.

The modern definition of a probability starts with a finite or countable set, the sample space, which relates to the set of all possible outcomes in a classical sense, denoted by Ω. It is then assumed that for each element $x \in \Omega$, an intrinsic probability value $f(x)$ is attached, which satisfies the following properties:

- $f(x) \in [0,1]$ for all $x \in \Omega$, and
- $\sum_{x \in \Omega} f(x) = 1$.

That is, the probability function $f(x)$ lies between zero and one for every value of x in the sample space Ω, and the sum of $f(x)$ over all values x in the sample space Ω is equal to 1. An event is defined as any subset E of the sample space Ω. The probability of the event E is defined as

$$P(E) = \sum_{x \in \Omega} f(x).$$

So, the probability of the entire sample space is 1, and the probability of the null event is 0. Continuous probability theory deals with events that occur in a continuous sample space. If the outcome space of a random variable X is the set of real numbers ($\mathcal{R}$) or a subset thereof, then a function called the cumulative distribution function F exists, defined by $F(x) = P(X \leqslant x)$; that is, $F(x)$ gives the probability that X will be less than or equal to x.

As F is continuous, its derivative exists and integrating the derivative gives us the cumulative distribution function; then the random variable X is said to have a probability density function (pdf), or more simply a density, $f(x) = dF(x)/dx$.

For a set $E \subseteq \mathcal{R}$, the probability of the random variable X being in E is

$$P(X \in E) = \int_{x \in E} dF(x),$$

this can be written as

$$P(X \in E) = \int_{x \in E} f(x)\, dx.$$

Whereas the pdf exists only for continuous random variables, the cumulative distribution function exists for all random variables (including discrete random variables) that take values in $\mathcal{R}$.

Intuition tells us that if a fair coin is tossed many times, then roughly half of the time it will turn up heads, and the other half it will turn up tails. Furthermore, the more often the coin is tossed, the more likely it should be that the ratio of the number of heads to the number of tails will approach unity. Probability theory provides a formal version of this intuitive idea, known as the law of large numbers. The law of large numbers states that the sample average

$$X_n = (1/n) \sum_{k=1}^{n} X_k$$

of a sequence of independent and identically distributed random variables X_k converges towards their common expectation, provided that the expectation of $|X_k|$ is finite.

The central limit theorem explains the ubiquitous occurrence of the normal distribution in nature. The theorem states that the average of many independent and identically distributed random variables with finite variance tends towards a normal distribution irrespective of the distribution followed by the original random variables. Formally, let $X_1, X_2,...$ be independent random variables with mean μ and variance $\sigma^2 > 0$. Then the sequence of random variables

$$Z_n = \sum_{i=1}^{n} (X_i - \mu)/\sigma\sqrt{n}$$

converges in a distribution of a standard normal random variable.

5.7 Expected value

In probability theory, the expected value of a random variable is, intuitively, the long-run average value of repetitions of an experiment; for example, the expected value in rolling a six-sided die is 3.5. More precisely, the law of large numbers states that the arithmetic mean of the values converges to the expected value as the number of repetitions becomes large.

The expected value of a discrete random variable is the probability-weighted average of all possible values. In other words, each possible value the random variable may assume is multiplied by its probability of occurring, and the resulting products are summed to produce the expected value. The same principle applies to a continuous random variable, except that an integral of the variable with respect to its probability density replaces the sum. For random variables, the long-tails of the distribution that represent such events prevent the summation from converging.

The expected value is a key aspect of how one characterizes a probability distribution. By contrast, the variance is a measure of dispersion of the possible values of the random variable around the expected value. The variance is the expected value of the squared deviation of the variable's value from the variable's expected value.

The expected value plays an important role in a variety of contexts: in regression analysis, one desires a formula or model to represent the observed data that gives a good estimate of the parameter(s) generating the effect of interest. The formula will give different estimates using different sets of data, so the estimate it gives is itself a random variable. A formula is considered good in this context if it is an unbiased estimator; that is, if the expected value of the estimate (the average value it would give over an arbitrarily large number of separate samples) can be shown to equal the true value of the desired parameter.

If the probability distribution of X admits a probability density function $f(x)$, then the expected value may be calculated as

$$E[X] = \int_{-\infty}^{\infty} x f(x) \, dx.$$

It is possible to construct an expected value equal to the probability of an event by taking the expectation of an indicator function that is unity, if the event has occurred, and zero otherwise. This relationship can be used to translate properties of expected values into properties of probabilities; for example, using the law of large numbers to justify estimating probabilities by frequencies.

The expected values of the powers of X are called the moments of X; the moments about the mean of X are expected values of powers of $X - E[X]$. The moments of some random variables can be used to specify their distributions, via appropriate moment generating functions.

To estimate the expected value of a random variable, one could repeatedly measure the variable and compute the arithmetic mean of the results. If the expected value exists, this procedure estimates the true expected value in an unbiased manner and has the advantage of minimizing the sum of the squares of the residuals (the sum of the squared differences between the observations and the estimate). The law of large numbers demonstrates that, as the size of the sample gets larger, the variance of this estimate gets smaller.

This property is often exploited in a wide variety of applications, including general problems of statistical estimation and machine learning, to estimate (probabilistically) quantities of interest via; for example, Monte Carlo methods, since most quantities of interest can be written in terms of an expectation; for example, $P(X \in A) = E[I_A(X)]$ where $I_A(X)$ is the indicator function for set A; that is, $X \in A \rightarrow I_A(X) = 1$, and $X \notin A \rightarrow I_A(X) = 0$.

In classical mechanics, the centre-of-mass is an analogous concept to expectation (as are the electrical moments of molecules, which characterize the charge distribution that is the molecule); for example, suppose X is a discrete random variable with values x_i and corresponding probabilities p_i. Now consider a weightless rod on which are placed weights, at locations x_i along the rod and having masses p_i (whose sum is one). The point at which the rod balances is $E[X]$; the mass of a probability distribution is balanced at the expected value.

Expected values can also be used to compute the variance, by means of the computational expression for the variance,

$$\text{Variance}(X) = E[X^2] - \left(E[X]\right)^2.$$

An important application of the expectation value is in quantum mechanics. The expectation value of a quantum mechanical operator A, operating on a quantum state defined by state vector $|\psi\rangle$ is written as $\langle A \rangle = \langle \psi |A| \psi \rangle$ where $\langle \psi | = | \psi \rangle^*$. The uncertainty in A can be calculated using the formula $(\Delta A)^2 = \langle A^2 \rangle - \langle A \rangle^2$.

Further reading

As stated earlier, the subject of measurement uncertainty is vast, but there are many standard textbooks. Some of these texts (or at least parts of them) are available on the Internet, Google them. In addition, there are many useful articles in the online encyclopaedia Wikipedia; for example, see the entries for measurement uncertainty, probability theory, and expected values. There are also freely available documents on measurement uncertainty on the websites the BIPM (www.bipm.org) listed under the heading publications.

Chapter 6

Guide to the Expression of Uncertainty in Measurement (the GUM)

6.1 Introduction

As we have seen, a measurement result is complete only when accompanied by a quantitative statement of its uncertainty. This requirement for a statement of uncertainty enables one to decide if the result is adequate for its intended purpose, and to determine if it is consistent with similar results. This qualification to a measurement may seem excessive, but is necessitated by the increasing dependence of the global economy upon technical measurements.

In the economic boom that followed the Second World War, it became apparent that the levels of precision seen in the various national industries (national measurement capacities), and the uncertainties associated with those levels of precision, were a fundamental factor in international trade and commerce. If a nation wished to increase its share of global trade it had to demonstrate quantitatively that it had the capacity to make the things needed by other nations, but at a unit cost and level of precision that was advantageously competitive. This was also the requirement that forced the UK and the USA to finally abandon their own customary weights and measures, and fully embrace the Metric System.

Over the years, many different approaches to evaluating and expressing the uncertainty of the results of measurements have been developed and used by different groups of scientists and different professions. But because of the difficulty in achieving international consensus on the expression of uncertainty in measurement, in 1977 the International Committee for Weights and Measures (CIPM, *Comité international des poids et mesures*), the world's highest authority in the field of measurement science, asked the BIPM (*Bureau international des poids et mesures*), an international agency created under the Metre Convention of 1875, to address this problem of finding an appropriate means of expressing uncertainty in measurement. This project was undertaken in collaboration with other international

doi:10.1088/978-1-6817-4433-9ch6

agencies and the various national measurement laboratories, and it led to the development of a recommendation or statement as to what constituted measurement uncertainty for consideration by the international physical-science community.

The uncertainty in the result of a measurement generally consists of several components, which may be grouped into two categories according to the manner in which their numerical value is estimated. As we saw earlier, this may be stated as:

- Type A uncertainty: those components which are evaluated by statistical methods.
- Type B uncertainty: those components which are evaluated by other means.

However, it is often difficult to find a simple correspondence between the classification into categories A or B. Any detailed report of uncertainty should consist of a complete list of all the components contributing to the uncertainty, specifying for each the method used to obtain its numerical value.

The components in category A are characterized by the estimated variances σ_i^2 (or the estimated standard deviation σ_i) and the number of degrees of freedom ν_i. Where appropriate the covariances should be given, which requires a statement of how the measurand is coupled to other properties of the system under study. The components in category B should be characterised by quantities u_j^2, which may be considered approximations to the corresponding variances, the existence of which is assumed. The quantities u_j^2 may be manipulated as if they were variances, and the quantities u_j treated as if they were standard deviations. Where appropriate, the covariances should be treated in a similar way.

The combined uncertainty should be characterized by the numerical value obtained by applying the usual method for the combination of variances (as outlined previously in the Propagation of uncertainty). The combined uncertainty and its components should be expressed in the form of standard deviations.

The above recommendations or suggestions are a brief outline rather than a detailed prescription. Subsequently, the CIPM worked with a wide-range of organizations with responsibility for the international coordination of technology, to develop a guidebook to the expression of uncertainty in measurements. This project developed into the *Guide to the Expression of Uncertainty in Measurement* (GUM), which is a summary of the methods of evaluating and expressing uncertainty in measurement, which has been widely adopted by the industries of many nations, by national metrology laboratories and many international organizations (see section 6.7 for more details).

6.2 Basic definitions

Measurement equation: the case of interest is where the quantity Y being measured, the measurand, is not measured directly but is determined from N other quantities $X_1, X_2, ..., X_N$ through a functional relation f, the measurement equation:

$$Y = f(X_1, X_2, ..., X_N). \tag{6.1}$$

Included among the quantities X_i are corrections (or correction factors), as well as quantities that take into account other sources of variability, such as different experimenters, different instruments, other samples, other laboratories, and different times at which observations are made. Thus, the function $f(X_1, X_2, ..., X_N)$ should express not simply a physical law, but a measurement process and, in particular, it should contain all the quantities that can contribute a significant uncertainty to the final measurement result.

An estimate of the measurand or output quantity Y, denoted by y, is obtained from equation (6.1) using input estimates $x_1, x_2, ..., x_N$ for the values of the N input quantities $X_1, X_2, ..., X_N$. Thus, the output estimate y, which is the result of the measurement, is given by

$$y = f(x_1, x_2, ..., x_N). \tag{6.2}$$

For example, if a voltage or potential difference V is applied to the terminals of a temperature-dependent resistor that has a resistance R_0 at the defined temperature t_0 and a linear temperature coefficient of resistance b, the power P (the measurand) dissipated by the resistor at the temperature t depends on V, R_0, b, and t according to:

$$P = f(V, R_0, b, t) = V^2/R_0\big[1 + b(t - t_0)\big]. \tag{6.3}$$

The uncertainty of the measurement result y arises from the uncertainties $u(x_i)$ (or u_i for simplicity) of the input estimates x_i that enter equation (6.2). Thus, in the example of equation (6.3), the uncertainty of the estimated value of the power P comes from the uncertainties of the estimated values of the potential difference V, resistance R_0, temperature coefficient of resistance b, and temperature t. In general, components of uncertainty may be categorized according to the method used to evaluate these quantities. With a Type A evaluation involving statistical analysis of series of observations, and a Type B evaluation involving evaluation by means other than statistical analysis of series of observations.

Each component of uncertainty, irrespective of how it has been evaluated, is represented by an estimated standard deviation, also termed standard uncertainty often with symbol u_i, and equal to the positive square root of the estimated variance. Thus, an uncertainty component obtained by a Type A evaluation is represented by a statistically estimated standard deviation σ_i, equal to the positive square root of the statistically estimated variance σ_i^2, and the associated number of degrees of freedom ν_i. For such a component, the standard uncertainty is $u_i = \sigma_i$. In a similar manner, an uncertainty component obtained by a Type B evaluation is represented by a quantity u_j, which may be considered an approximation to the corresponding standard deviation; it is equal to the positive square root of u_j^2, which may be considered an approximation to the corresponding variance and which is obtained from an assumed probability distribution based on all available information. Since the quantity u_j^2 is treated like a variance and u_j like a standard deviation, for such a component the standard uncertainty is simply u_j.

6.3 Evaluating uncertainty components

A Type A evaluation of standard uncertainty may be based on any valid statistical method for treating data; for example, calculating the standard deviation of the mean of a series of independent observations; using the method of least-squares to fit a curve to data in order to estimate the parameters of the curve and their standard deviations; and carrying out an analysis of variance (ANOVA)[1] in order to identify and quantify random effects in certain kinds of measurements.

In the case where an input quantity X takes random values from a finite data set $x_1, x_2, ..., x_N$, with each value having the same probability, the standard deviation and the mean (μ) are:

$$\sigma = \sqrt{\frac{1}{N}\left[(x_1 - \mu)^2 + (x_2 - \mu)^2 + \cdots + (x_N - \mu)^2\right]}, \quad \text{where}$$

$$\mu = \frac{1}{N}(x_1 + \cdots + x_N), \tag{6.4}$$

or,

$$\sigma = \sqrt{\frac{1}{N}\sum_{i=1}^{N}(x_i - \mu)^2}, \quad \text{where} \quad \mu = \frac{1}{N}\sum_{i=1}^{N}x_i. \tag{6.5}$$

If, instead of having equal probabilities, the values have different probabilities, let x_1 have probability p_1, x_2 have probability p_2, ..., x_N have probability p_N. In this case, the standard deviation will be

$$\sigma = \sqrt{\sum_{i=1}^{N}p_i(x_i - \mu)^2}, \quad \text{where} \quad \mu = \sum_{i=1}^{N}p_i x_i. \tag{6.6}$$

If it had been possible, this would have been the appropriate place for Pierre Méchain and his surveying colleague to have carefully considered which of their data they would have preferred to keep in their final analysis of the metric survey, and which data they would have chosen to exclude from the final calculation. Analysis of Type B uncertainty allows one to put a subjective analysis of the quality of measurements onto a firmer quantitative basis. It turns an art into a science. A Type B evaluation of standard uncertainty should be based on scientific judgment using all of the relevant information available; for example,

---

[1] Analysis of variance (ANOVA) is a set of statistical models used to analyze the differences among group means (such as variation among and between groups), developed by the British statistician and evolutionary biologist Ronald Fisher (1890–1962). In an analysis by ANOVA, the observed variance in a particular variable is split into components attributable to different sources of variation. In its simplest form, ANOVA provides a statistical test of whether or not the means of several groups are equal, and therefore generalizes the t-test to more than two groups. Such an analysis is useful for comparing three or more means (of groups) for statistical significance. It is conceptually similar to multiple two-sample t-tests, but is more conservative (results in less type I hypothesis testing errors) and is therefore suited to a wide range of practical problems; see chapter 7 on clinical trials.

- previous measurement data,
- experience with, or general knowledge of, the behaviour and property of relevant materials and instruments,
- manufacturer's specifications,
- data provided in calibration and other reports, and
- uncertainties assigned to reference data taken from handbooks.

Below are some examples of Type B evaluations in different situations, depending on the available information and the assumptions of the experimenter. Generally, the uncertainty is derived either from an outside source, or calculated from an assumed distribution.

6.4 Uncertainty derived from some assumed distibutions[2]

Normal distribution: if the quantity of interest is modelled by a normal or Gaussian probability distribution, there are no finite limits that will contain 100% of its possible values. However, plus and minus three standard deviations about the mean of a normal distribution corresponds to 99.73% confidence limits. Thus, if the lower and upper limits, a_- and a_+, respectively, of a normally distributed quantity with mean $(a_+ + a_-)/2$ are considered to contain almost all of the possible values of the quantity; that is, 99.73% of them, then u_j is approximately $a/3$, where $a = (a_+ - a_-)/2$ is the half-width of the interval.

Uniform (rectangular) distribution: Estimate lower and upper limits a_- and a_+, respectively, for the value of the input quantity such that the probability that the value lies in the interval a_- to a_+ is 100%. Provided that there is no contradictory information, treat the quantity as if it is equally probable for its value to lie anywhere within the interval a_- to a_+; that is, model it by a uniform (that is, rectangular) probability distribution. The best estimate of the value of the quantity is then $(a_+ + a_-)/2$ with $u_j = a/\sqrt{3}$, where $a = (a_+ - a_-)/2$ is the half-width of the interval.

Triangular distribution: The rectangular distribution is a reasonable default model in the absence of any information about the distribution of the quantity of interest. But if it is known that values of the quantity in question near the centre of the limits are more likely than values close to the limits, a normal distribution or, for simplicity, a triangular distribution, may be a better and simpler model. For a triangular distribution, estimate lower and upper limits a_- and a_+ for the value of the input quantity in question, such that the probability that the value lies in the interval a_- to a_+ is 100%. Provided that there is no contradictory information, model the quantity by a triangular probability distribution. The best estimate of the value of the quantity is then $(a_+ + a_-)/2$ with $u_j = a/\sqrt{6}$, where $a = (a_+ - a_-)/2$ is the half-width of the interval.

---

[2] For further details see http://physics.nist.gov/cuu/Uncertainty/typeb.html.

6.5 Combining uncertainty components

The combined standard uncertainty of the measurement result y, designated by $u_c(y)$ and taken to represent the estimated standard deviation of the result, is the positive square root of the estimated variance $u_c^2(y)$ obtained from:

$$u_c^2(y) = \sum_{i=1}^{N} \left(\frac{\partial f}{\partial x_i}\right)^2 u^2(x_i) + 2\sum_{i=1}^{N-1}\sum_{j=i+1}^{N} \frac{\partial f}{\partial x_i}\frac{\partial f}{\partial x_j}u(x_i, x_j) \tag{6.7}$$

Equation (6.7) is based on a first-order Taylor series approximation of the measurement equation $Y = f(X_1, X_2, ..., X_N)$ given in equation (6.1), and is referred to as the law of propagation of uncertainty (see previous chapter). The partial derivatives of f with respect to X_i (sometimes referred to as sensitivity coefficients) are equal to the partial derivatives of f with respect to X_i evaluated at $X_i = x_i$; and $u(x_i)$ is the standard uncertainty associated with the input estimate x_i; and $u(x_i, x_j)$ is the estimated covariance associated with x_i and x_j. Equation (6.7) is often reduced to a simple form; for example, if the input estimates x_i of the input quantities X_i can be assumed to be uncorrelated, then the second term vanishes. Further, if the input estimates are uncorrelated and the measurement equation is one of the following two forms, then equation (6.7) becomes simpler still.

1 Assuming a measurement equation to consist of a sum of quantities X_i multiplied by constants a_i; that is, $Y = a_1 X_1 + a_2 X_2 + \cdots + a_N X_N$, and with measurement result: $y = a_1 x_1 + a_2 x_2 + \cdots + a_N x_N$, the combined standard uncertainty can be written as: $u_c^2(y) = a_1^2 u^2(x_1) + a_2^2 u^2(x_2) + \cdots a_N^2 u^2(x_N)$.

2 If the measurement equation is a product of quantities X_i, raised to powers a, b, ... p, multiplied by a constant A; that is, $Y = A X_1^a X_2^b ... X_N^p$, and with measurement result: $y = A x_1^a x_2^b ... x_N^p$, the combined standard uncertainty can be written as: $u_{c,r}^2(y) = a^2 u_r^2(x_1) + b^2 u_r^2(x_2) + \cdots + p^2 u_r^2(x_N)$.

Here $u_r(x_i)$ is the relative standard uncertainty of x_i and is defined by $u_r(x_i) = u(x_i)/|x_i|$, where $|x_i|$ is the absolute value of x_i and x_i is not equal to zero; and $u_{c,r}(y)$ is the relative combined standard uncertainty of y and is defined by $u_{c,r}(y) = u_c(y)/|y|$, where $|y|$ is the absolute value of y and y is not equal to zero.

If the probability distribution characterized by the measurement result y and its combined standard uncertainty $u_c(y)$ is approximately normal, and $u_c(y)$ is a reliable estimate of the standard deviation of y, then the interval $y + u_c(y)$ to $y - u_c(y)$ is expected to encompass approximately 68% of the distribution of values that could reasonably be attributed to the value of the quantity Y for which y is an estimate. This implies that we can state that with a level of confidence of 68%, Y is greater than or equal to $y - u_c(y)$, and is less than or equal to $y + u_c(y)$, or $Y = y \pm u_c(y)$.

6.6 Expanded uncertainty and coverage factor

Although the combined standard uncertainty u_c is used to express the uncertainty of many measurement results, for some commercial, industrial, and regulatory applications (e.g. when health and safety are concerned), what is often required is a measure of uncertainty that defines an interval about the measurement result y, within which the value of the measurand Y is believed to lie with a defined certainty. The measure of uncertainty intended to meet this requirement is termed expanded uncertainty, symbol U, and is obtained by multiplying $u_c(y)$ by a coverage factor, k. Thus $U = ku_c(y)$ and it is confidently believed that Y is greater than or equal to $y - U$, and is less than or equal to $y + U$, or $Y = y \pm U$.

In general, the value of the coverage factor k is chosen on the basis of the desired level of confidence to be associated with the interval defined by $U = ku_c$. Typically, k is in the range two to three. When the normal distribution applies and u_c is a reliable estimate of the standard deviation of y, $U = 2\,u_c$ (i.e. $k = 2$) defines an interval having a level of confidence of approximately 95%, and $U = 3\,u_c$ (i.e. $k = 3$) defines an interval having a level of confidence greater than 99%.

The GUM is an entire way of looking at, and treating measurement uncertainty, and represents over thirty years of research and development, and of international collaboration. The principles and concepts underlying the GUM include the following:

- A measurement model relating functionally one or more output quantities, about which we are seeking information, to input quantities, about which we possess information.
- Modelling our knowledge about the measurement of a quantity in terms of a probability distribution.
- Estimate the expectation and standard deviation (standard uncertainty) of a quantity characterized by a probability distribution.
- The use of new information to update an input probability density function: Bayes' Theorem.
- Assignment of a probability density function to a quantity using the Principle of Maximum (information) Entropy.
- Determination of the distribution for an output quantity (or the joint distribution for more than one output quantity) using the propagation of distributions.

Figure 6.1 summarizes the principles of the GUM methodology. The model equation provides the basis for the propagation of the probability density functions for the input quantities, and in the case of using Gaussian uncertainty propagation, for the propagation of their expectation values and associated uncertainties. Consequently, it is the modelling of the measurement process that is the key element of modern uncertainty evaluation, irrespective of the method used for calculating the uncertainty.

The GUM is the 'gold standard' of estimating uncertainty in measurements; particularly, measurements made at an international level and which have the backing of law; for example, defining and maintaining the world standard for mass using the data displayed in figure 4.2 in section 4.2 or to define the precision and uncertainty of international atomic time (UTC or *Temps universel coordonné*). An

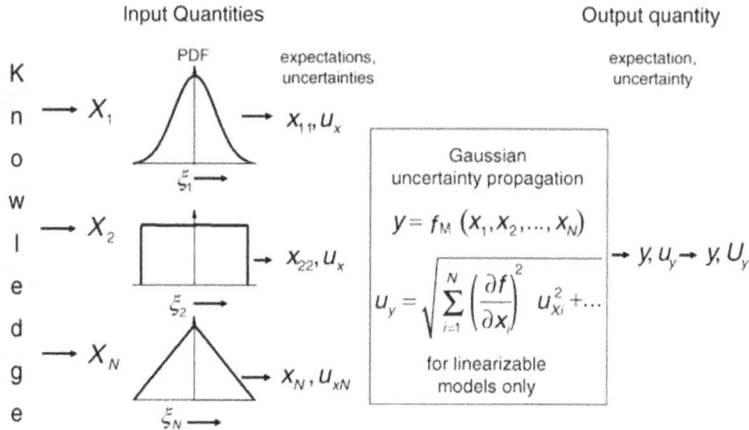

Figure 6.1. Illustration of the GUM procedure for the expression of measurement uncertainty. Symbols: Y is the measurand, $X_1 \ldots X_N$ are the input quantities, $y = E[Y]$ is the expectation of the probability density function for the measurand (taken as a best estimate), u_y is the standard uncertainty associated with y, $x_1 \ldots x_N$ are the expectations for the probability density functions for the input quantities, $u_{x1} \ldots u_{xN}$ are the standard uncertainties associated with $x_1 \ldots x_N$, U_y is the expanded measurement uncertainty. (Image taken from Sommer K D and Siebert B R L 2006 *Metrologia* **43** S200–S210; thanks to the BIPM for permission to reproduce this image.)

ordinary research student working away in a small university laboratory is not seeking such an exhaustive formulation for his or her estimation of uncertainty, but the general principles of the GUM are good practice and worth following. After all, every scientist is in some way a metrologist, we all make measurements and are trained to look at the world with an experimenter's eye; unless, of course, you prefer to do theory.

Further reading

There are many freely available Internet sources where documentation about the GUM and the application of the GUM to specific problems may be found; for example, the BIPM website www.bipm.org/en/publications/guides/gum.html provides access to a range of documents relating to the GUM (including the main document running to 134 pages including references), and the evolution of the GUM; there is also the NIST (National Institute of Standards and Technology) website at http://www.nist.gov where one may use the search facility to find documents relating to measurement uncertainty and to the GUM. In addition, at www.bipm.org/en/publications/guides/vim.html, you can also find the international vocabulary of metrology; and international and US perspectives on measurement uncertainty (http://www.physics.nist.gov/cuu/Uncertainty/international1.html) is also well worth investigating.

In addition, there are many entries on the online encyclopaedia, Wikipedia that are well worth reading; for example, the entry on standard deviation.

If the reader is interested in the GUM, I would also like to draw their attention to a special issue of the journal *Metrologia* (which is published by IOP Publishing on behalf of the BIPM) on Statistical and Probabilistic Methods for Metrology (volume 43, issue number **4**, 2006). This special issue contains a great many technical articles on the applications of the GUM; in particular, K D Sommer and B R L Siebert 2006 Systematic approach to the modelling of measurements for uncertainty evaluation *Metrologia* **43** S200.

Chapter 7

Clinical trials

Statistical thinking will one day be as necessary for efficient citizenship as the ability to read and write.

H G Wells, 1866–1946

7.1 Introduction

Clinical research includes investigating the efficacy and long-term benefit of a proposed treatment, assessing the relative benefits of competing therapies, and establishing optimal treatment combinations. Such research attempts to answer questions such as: should a man with prostate cancer undergo surgery or receive radiation, or should the clinicians merely wait watchfully? Then there are the questions about new drugs; for example, is the incidence of serious adverse effects among patients receiving a new pain-relieving therapy greater than the incidence of serious adverse effects in patients receiving the standard therapy, and how does a new anti-HIV molecule compare with those presently in use when used in combination with drugs A and B?

Before the widespread use of experimental trials, clinicians attempted to answer such questions by generalizing from their experiences with individual patients to the population at large. Clinical judgement and reasoning were applied to reports of interesting cases in the pursuit of new, best medical practice. The concept of variability among individuals and its sources may have been noted, but were not addressed formally.

In the 20th Century, the applications of statistics evolved rapidly and it began to be applied to clinical research. However, the statistical approach used in clinical research is different from that used in physics; in medicine, one is more interested in the theoretical science or formal study of the inferential process, especially the planning and analysis of experiments, surveys, and observational studies. Statistical methods provide formal accounting for variability in patients' responses to treatment; that is, statistics allow the clinicians to put on a firm quantitative basis the wide-range of responses seen in a collection of patients, when they are all given the same treatment. The use of statistics allows the clinical researcher to form reasonable and accurate inferences from collected information, and to make sound decisions in the presence of uncertainty.

Clinical and statistical reasoning are both crucial to progress in medicine. Clinical researchers must generalize from the few to the many, and combine empirical evidence with theory. In both medical and statistical sciences, empirical knowledge is generated from observations and data, and as we have seen, there is no data without uncertainty (and as we saw in chapter 4, there is generally a large uncertainty attached to automated measurements made in a hospital laboratory). Medical theory is based upon established biology and hypotheses. To establish a hypothesis requires both a theoretical basis in biology and statistical support for the hypothesis, based on the observed data and a theoretical statistical model.

An experiment may be thought of as a series of observations made under conditions controlled by a scientist. A clinical trial is an experiment testing medical treatments on human subjects. The clinical investigator controls factors that contribute to variability and bias such as the selection of subjects, application of the treatment, evaluation of outcome, and methods of analysis. The distinction of a clinical trial from other types of medical studies is the experimental nature of the trial, and the use of human trial subjects. The term clinical trial is preferred over clinical experiment, because the latter may suggest disrespect for the value of human life. But whatever it is called, it is a study that can last a long time.

The reason clinical trials are long is that clinicians are interested in the long-term (that is, over at least five years) benefits or otherwise of; for example, a new drug, together with the advent and severity of any associated side-effects. The trial may also be very large; that is, there may be over 2000 patients enrolled in the trial as one needs to investigate a significant, representative sample of humanity. Whereas all hydrogen atoms or all carbon dioxide molecules will behave in exactly the same way when stimulated in the same manner, the same is not true of humans. No two humans are alike in their biological response to medication, and men behave differently from women. So the most important question asked by those designing a clinical trial is about sample size. That is, how many individuals are needed to mask the genetic variability and individual uniqueness that has arisen through millions of years of evolution? This individual uniqueness can be considered to be the noise and uncertainty against which the clinicians are seeking to observe the effect they are after. No two people respond in the same way to the same size dose of drug X; so how many trial participants are required to average out natural genetic variability. After this, the question will be the percentage of men and women and the age groups to be studied; this is where the statistics is required, in trying to identify meaningful results in what could be a long, expensive trial/experiment.

Design is therefore the process or structure that isolates the factors of interest. Although the researcher designs a trial to control variability due to factors other than the treatment of interest, there is inherently larger variability in research involving humans than in a controlled laboratory situation. In many ways the design of a study is more important than the analysis. A badly designed trial can never be retrieved, whereas a poorly analysed one can usually be re-analyzed. Consideration of design is also important because the design of a study will govern how the data are to be analysed.

Most medical studies consider an input, which may be a medical intervention or exposure to a potentially toxic new drug, and an output, which is some measure of health that the intervention is supposed to affect. The data that will be analysed will resemble the data given in figure 4.3 in section 4.3 but for many individuals. The simplest way to categorize studies is with reference to the time sequence in which the input and output are studied. The most powerful studies are prospective studies where the paradigm is the randomized controlled trial. Here, subjects with a disease are randomized to one of two (or more) treatments, one of which may be a control treatment. The importance of randomization is that, in the long run, the clinicians know that treatment groups will be balanced in known and unknown prognostic factors. It is also important that the treatments are concurrent—that the active and control treatments occur in the same period of time.

One of the major threats to the validity of a clinical trial, and a major source of uncertainty, is compliance. Patients are likely to drop out of trials if the treatment is unpleasant, and often fail to take medication as prescribed. It is usual to adopt a pragmatic approach and analyse by intention to treat; that is, analyse the study by the treatment that the subject was assigned to, not the one they actually took.

7.2 Sample size

As mentioned earlier, this is the most important question in the design of a trial. So one of the most frequently asked questions put to a statistician about the design of the trial is the number of patients to be included. It is an important question, because if a study is too small it will not be able to answer the question posed, and would be a waste of effort. It could also be deemed unethical because patients may be put at risk with no apparent benefit. However, studies should not be too large because resources would be wasted if fewer patients would have sufficed. The sample size depends on four critical quantities: the type I and type II error rates α and β (see below), the variability of the data σ^2, and the effect size, d.[1] In a clinical trial, the effect size is the amount by which we would expect the two treatments to differ, or is the difference that would be clinically worthwhile.

Usually α and β are fixed at 5% and 20% (or 10%), respectively. A simple formula for a two group, parallel trial with a continuous outcome is that the required sample size per group is given by $n = 16 \, \sigma^2/d^2$ for each of the two sides, with α of 5% and β of 20%, respectively. For example, in a trial to reduce blood pressure, if a clinically worthwhile effect for diastolic blood pressure is a reduction of 5 mm Hg and

---

[1] The type I and type II error rates mentioned here have little to do with the Type A and Type B uncertainties we encountered earlier in this volume when looking at the analysis of measurement uncertainty. There is still a gulf between the terminology, vocabulary, procedures and definitions used by clinical statisticians and metrologists. These terms are also used in a more general way by social scientists to refer to flaws in reasoning.

the observed effect between subjects (the standard deviation) is 10 mm Hg; that is, 5 ± 5 mm Hg, and one can immediately see the difference between statistics in clinical trials and in physics experiments. This is a difference which comes essentially from the genetic variability of the subjects in the clinical trials. In such a trial, we would require $n = 16 \times 100/25 = 64$ patients per group to try and mask these natural variations, and lead to a useful result. The sample size goes up as the square of the standard deviation of the data (the variance) and goes down inversely as the square of the effect size. Doubling the effect size (looking for a bigger effect) reduces the sample size by a factor four, it is much easier to detect large effects. In practice, the sample size is often fixed by other criteria, such as finance or resources, and the formula is used to determine a realistic effect size.

7.3 Statistical hypothesis testing

As mentioned in the above footnote, the basis of the statistical approach to clinical research is different from the approach adopted in the GUM. The clinical researcher is interested in testing hypotheses and in analyzing the results of their trials to see if the hypothesis has been upheld or not. And so there is little connection between the Type A and Type B uncertainties of the GUM, and the type I and type II error of statistical hypothesis testing. Hypothesis testing does of course take place in physics and chemistry, but usually when a researcher is trying to observe or measure a phenomenon for the first time. The vocabularies used by clinical researchers and metrologists are different, yet they are both based on the same scientific principles, thereby demonstrating the ongoing need for rationalization of definitions and nomenclature.

In statistics, a null hypothesis is a statement that one seeks to nullify with evidence to the contrary. Most commonly it is a statement that the phenomenon being studied produces no effect or makes no difference; for example, 'this diet has no effect on people's weight'. Usually, an experimenter frames a null hypothesis with the intent of rejecting it; that is, intending to run an experiment which produces data that shows that the phenomenon under study does make a difference. In some cases there is a specific alternative hypothesis that is opposed to the null hypothesis, in other cases the alternative hypothesis is not explicitly stated, or is simply 'the null hypothesis is false'. In either event, this is a binary judgment, but the interpretation differs, and is a matter of dispute between the various schools of statistics.

A type I error, or error of the first kind is the incorrect rejection of a true null hypothesis. Usually a type I error leads one to conclude that a supposed effect or relationship exists when in fact it doesn't. Examples of type I errors include a test that shows a patient to have a disease when in fact the patient does not have the disease, a fire alarm going off indicating a fire when in fact there is no fire, or an experiment indicating that a medical treatment should cure a disease when in fact it does not.

A type II error, or error of the second kind is the failure to reject a false null hypothesis. Examples of type II errors would be a blood test failing to detect the disease it was designed to detect in a patient who has the disease; a fire breaking out and the fire alarm not ringing; or a clinical trial of a treatment failing to show that the treatment works when in fact it does.

In terms of false positives and false negatives, a positive result corresponds to rejecting the null hypothesis, while a negative result corresponds to failing to reject the null hypothesis. In these terms, a type I error is a false positive, and a type II error is a false negative.

When comparing two means of the results of two sets of data, and concluding that the means were different when in reality they were not different would be a type I error; concluding the means were not different when in reality they were different would be a type II error (see comments in section 4.3 about identifying real differences between successive routine blood measurements.

All statistical hypothesis tests have a probability of making type I and type II errors. For example, all blood tests for a disease will falsely detect the disease in some proportion of people who do not have the disease, and will fail to detect the disease in some proportion of people who do have it. There may be a perfectly good biological reason for this; perhaps, a set of rare genes produces a marker that triggers or nullifies the test in the same way that the vector of the disease would. This is a delicate situation requiring large groups in clinical trials to be tested fully. A test's probability of making a type I error is denoted by α. A test's probability of making a type II error is denoted by β. These error rates are traded off against each other; for any given sample set, the effort to reduce one type of error generally results in increasing the other type of error. For a given test, the only way to reduce both error rates is to increase the sample size, and this may not be feasible.

If we wished to test the hypothesis: a patient's symptoms improve after treatment A more rapidly than after a placebo treatment. The null hypothesis would be: a patient's symptoms after treatment A are indistinguishable from a placebo. The null hypothesis can be tested relatively straightforwardly, and if it is found to be the case that a patient's symptoms after treatment A are indistinguishable from a placebo (this is not as obvious as it may sound; remember the strange and unexplained 'placebo effect'), then one will need to investigate what exactly went on in the trial.

This kind of hypothesis error may happen due to staff failing to keep patients unaware of which treatment they're receiving, to uncontrolled variables, to difficulty tracking subjective symptoms like pain or for many other reasons. A type I error would falsely indicate that treatment A is substantially more effective than the placebo, whereas in a type II error the mistake is believing that treatment A has no effect. Statistical tests always involve a trade-off between:

- the acceptable level of false positives, and
- the acceptable level of false negatives.

The notions of false positives and false negatives is not limited to clinical research, but has a wide currency in the world of computers and security applications. Security and breeches of security are important considerations in keeping computer data safe, while maintaining access to that data for appropriate users. We need to consider:

- avoiding the type I errors (or false negatives) that classify authorized users as intruders, and

- avoiding the type II errors (or false positives) that classify imposters as authorized users.

False positives are routinely found every day in airport security screening, which are essentially visual inspection systems. The installed security alarms are intended to prevent weapons being brought onto aircraft, yet they are often set to such high sensitivity that the alarm goes off many times a day for minor items, such as keys, belt buckles, loose change, mobile phones, and even metal tacks stuck in the soles of shoes.

The ratio of false positives (identifying an innocent traveller as a terrorist) to true positives (detecting a would-be terrorist) is therefore high; and because almost every alarm is a false positive, the positive predictive value of these screening tests is low.

The relative cost of false results determines the likelihood that test creators allow these events to occur. As the cost of a false negative in this scenario is extremely high (not detecting a weapon or a bomb being brought onto a plane could result in hundreds of deaths) whilst the cost of a false positive is relatively low (further inspection) the most appropriate test is one with a low statistical specificity but high statistical sensitivity (one that allows a high rate of false positives in return for minimal false negatives).

Although it might be imagined that there is a close similarity between the theory of measurement uncertainty in the physical sciences and the statistical interpretation of clinical trials in medicine, the truth is that these two hugely important fields of research are far apart. They are far apart in nomenclature, and in the magnitude of the uncertainties being considered. And this difference in the magnitude of the uncertainties is the result of the inescapable uniqueness of individuals (their genetic variability), as opposed to the identical nature of all; for example, carbon dioxide molecules. In figure 4.1 in section 4.1, we see the behaviour of about 10^{24} molecules of carbon dioxide all doing the same thing when subjected to an applied electric-field gradient, whereas in figure 4.3, we see the data generated by a single person trying to survive.

It is the inevitable, large uncertainty in the biological behaviour of a group of individuals (the genetic variability, which is the result of millions of years of evolution) that does not permit a GUM or even a GUM-like approach to quantifying uncertainty in clinical trials. So, we have to use statistical inference, with its larger uncertainties, when we come to dealing with the behaviour (even the biochemical and biological responses) of groups of people. Consider the data in figure 4.3(b), where we have an uncertainty on each measurement of $\pm50\%$; then imagine a statistical analysis of the results of a clinical trial involving many thousands of people being similarly treated. The uncertainties would become truly large, and perhaps meaninglessly large, as far as a physicist would be concerned, but such patients must still be treated and well treated. In physics, there is generally greater precision and smaller uncertainty on a measurement, and so sophisticated models of expressing that uncertainty may be adopted.

Chapter 8

Direct measurements: quadrupole moments and stray light levels

To call in the statistician after the experiment is done may be no more than asking him to perform a post-mortem examination: he may be able to say what the experiment died of.

Ronald Aylmer Fisher, 1890–1962

8.1 Introduction

As we know, not all atoms are alike. Atoms such as fluorine and chlorine have a propensity, when covalently bonded with other atoms for attracting an additional electron into their structure and becoming slightly negatively-charged. This attractive pull for electrons is termed the atom's electronegativity (a heuristic quantity developed in chemistry by the American chemist and double Nobel Laureate, Linus Pauling, 1901–1994). Fluorine has the highest electronegativity and hydrogen has a low electronegativity; that is, a fluorine atom when covalently bonded to another atom will pull an electron from the other atom towards itself, but hydrogen will be less able to pull an electron to itself, but will allow its single electron to be pulled closer to the more electronegative atom to which it may be bonded.

What this electronegativity means is that if a hydrogen atom and a fluorine atom combine chemically to form hydrogen fluoride (HF), although the new molecule will not have an overall electrical positive- or negative-charge, there will be a slight separation of the charge-density within the molecule. Chemists talk about the fluorine end of the molecule being slightly more negatively charged than the hydrogen end of the molecule, which will be slightly more positively charged ($\delta+$ H-F $\delta-$). However, the overall electrical charge will be zero, so the molecule will be uncharged. A symmetric molecule such as hydrogen (H_2) has no such disproportionate or unbalanced electronegativity, because of the symmetry.

doi:10.1088/978-1-6817-4433-9ch8

The consequence of such charge asymmetries in molecules is that there is a slight net attraction between any two such molecules. If one molecule has a slightly negatively-charged end and a slightly positively-charged end, then when two such molecules encounter each other there will be a slight electrostatic attraction with the slightly negative-end of one molecule seeking (for reasons of attractive electrostatic interaction) to be closer to the slightly positive-end of the other molecule. This type of weak electrostatic interaction is manifest in more energy being required to separate individual molecules from bulk quantities of those molecules; that is, the boiling temperature of HF is higher that the boiling temperatures of H_2 and of F_2.

If the charge distribution in a molecule is not perfectly spherical, as it is in an atom of helium, there will always be a small electrical moment in that molecule. These electrical moments are related to the shape or symmetry of the molecule, and as the molecule becomes increasingly symmetric; that is, on going from a linear shape to a tetrahedron and on to an octahedron and then to a sphere, the electrical moments become increasingly smaller and smaller; becoming zero in a sphere, the perfect Platonic body.

It was the Dutch-American physicist and Nobel Laureate, Peter Debye (1884–1966), who was the first to explain the nature and origin of the small electric dipole moments seen in molecules; that is, a separation of charge onto two centres in the molecule. Debye quantified the polarization in molecules and introduced the idea of a charge separation, defining in the process a new unit of measurement; one unit of electrical charge separated by one unit of length (that is, electric charge by distance). In addition, he developed a means of measuring the absolute value of molecular dipole moments, which is still in use today. The absolute value of the dipole moment is its size (or magnitude) and its sign (greater than zero or less than zero); if you know the sign of the dipole moment, you can determine which end is positively-charged and which end is negatively-charged.

The unit of distance Debye chose to work with when creating his unit for the dipole moment was the angstrom (a unit of length named after the Swedish physicist Anders Ångström; 1 Å equals 10^{-10} metre), because it was ideally suited to describing the size of the molecule; the bond between the hydrogen atom and the fluorine atom in hydrogen fluoride is about one angstrom in length. Debye defined a dipole moment as being one unit of electric charge separated by one angstrom; this convenient measure means that molecules have dipole moments of about one or two such units. This unit proved to be so useful that it was named, by Debye's colleagues, a Debye. One Debye (abbreviated as 1 D) is one unit of charge separated by one angstrom; in HF, the dipole moment is 1.86 D with the negative end on the fluorine atom.

In the HF molecule, the centres of gravity of the distribution of negative-charge and the centre of gravity of positive-charge do not coincide, as they would in a molecule of hydrogen (H_2). In such dipolar molecules, there are lines of force going from the positive-end of the molecule to the negative-end (this is a convention).

In the International System of Units (SI), a molecular dipole moment is defined in coulomb metres or C m (SI units of electric charge, the coulomb, and distance, the metre), and typical values are 3.33×10^{-30} C m, and so the SI value of the dipole moment of HF would be 6.19×10^{-30} C m. It is easier to tabulate, remember and discuss values of molecular properties which are, of order, one to five rather than something multiplied by 10^{-30}, which is the reason why there is still no single universal system of units in use by scientists; scientists tend to use whichever system of units is convenient for them.

In carbon dioxide, CO_2, which is linear with the carbon atom symmetrically placed between two more electronegative oxygen atoms, the two bond dipole moments are opposed (pointing in opposite directions) and the net dipole moment of the molecule is zero. However, the molecule is better thought of as being oval in shape and not infinitely narrow (that is, thought of as being 3-dimensional rather than 1-dimensional). In this way, we see that the CO_2 molecule is truly like a fat cigar with a finite width, and so even if the net dipole moment is zero, there is still an asymmetry in the charge distribution—there is a lot of charge-density on the long axis and some charge-density perpendicular to the long axis. This geometry defines an electric quadrupole moment. That is, the charge asymmetry in the molecule may be thought of as residing on four centres; at each end and at the middle, but off axis. So here we have a charge distributed over an area; that is, charge by distance by distance or in the SI system, coulomb metres $\times$ metres or C m^2, with values typically 3.33×10^{-40} C m^2.

The first direct method for measuring the absolute value (sign and magnitude) for a molecular quadrupole moment was devised in 1959 by A David Buckingham (born 1930) [1]. The technique suggested by Buckingham proved to be successful and Peter Debye suggested in, *Chemical and Engineering News*, the news magazine of the American Chemical Society [2] that the unit of the quadrupole moment be called the Buckingham (that is, one unit of charge distributed over one square angstrom be called one Buckingham), so the quadrupole moment of CO_2 would be about 4.5 Buckinghams or 15×10^{-40} C m^2 in the SI.[1]

After the quadrupole moment, the next highest electrical moment of a molecule is the octupole moment, which would be one unit of electric charge spread throughout a volume or one unit of electric charge in a cubic angstrom or about 3.33×10^{-50} C m^3 in the SI. (We are now starting to become seriously small.) Molecules possessing an octupole moment, arising from the asymmetry in the distribution of electronic charge, are tetrahedral in shape; for example, methane or carbon tetrachloride. The next highest electrical moment would the hexadecapole moment; such a molecule would be an octahedron, as in sulphur hexafluoride. These higher moments are very small, compared with the dipole moment of a small molecule such as water; they represent ever-smaller departures from a symmetric distribution of charge.

---

[1] There is an amusing corollary to this story of the naming of units. The author was a research student of Professor Buckingham, and one day, long-ago we were discussing this story of how the unit of the quadrupole moment came to be called the Buckingham (my thesis was essentially about the measurement of this molecular property), and my supervisor said to me with a smile, 'if you work out a way of measuring the octupole moment, then I will suggest that the unit be called the Williams'. To date, there is still no method for the determination of the sign and magnitude of the octupole moment.

8.2 Measuring the quadrupole moments of molecules

The experiment for measuring the molecular quadrupole moment of a molecule such as CO_2 or benzene is due to Buckingham [1]. Essentially, a large well-defined electric-field gradient is applied to a sample of the gas or vapour, of known number density, and a polarized light (laser) beam is passed through the medium in the presence of the applied electric-field gradient. If the laser light is initially plane-polarized, it will become elliptically-polarized on passage through the birefringent medium arising from the molecules oriented by the applied electric-field gradient.

This technique of measurement arises from the 19th Century observations of the Kerr effect in Glasgow in the 1870s. Under the influence of James Clerk Maxwell, John Kerr (who was a theology student) was seeking to observe the effect of an applied electric field on a beam of light. To undertake this experiment he needed to pass a beam of polarized light between the charged electrodes that generated the electric field. Of course, what he was also doing was observing the effect of the applied field on the polarization of the light beam as it passed through the medium (a dense, optically-pure lead glass) placed between the electrodes. What Kerr observed was that the initially linearly-polarized beam of light became elliptically polarized on passing through a medium subject to the applied electric field. That is, the electric field caused the medium under study to behave as a birefringent or uniaxial crystal such as calcite.

In the Kerr effect, there is no rotation of the plane of polarization of the light upon propagation through the material in the applied electric field, which would be dichroism, instead an additional plane of polarization is generated perpendicular to that of the incident light. That is, the material behaves as a birefringent or uniaxial crystal under the influence of the electric field; there are now two routes for the light to propagate through the glass.

The Kerr effect is a means of imposing order on a collection of molecules (just as the magnetic field of a permanent magnet will impose order on a sample of iron filings). The applied electric field imposes an ordered, crystal-like structure on an initially structure-less sample. In a liquid there is no permanent structural order, the molecules are tumbling over each other while keeping a constant separation, and the higher the temperature, the greater the tumbling motion and the degree of disorder. If an electric field is applied to a liquid, which possesses a large Kerr effect, the molecules become ordered by the interaction of the applied electric field and the electrical properties (in particular, the electric dipole moment) of the molecules; the molecules will line-up following the applied electric force-field. However, as this ordering effect is working against the effect of temperature, which is seeking to randomize any order, the Kerr effect is temperature-dependent; the lower the temperature, the bigger the observed Kerr effect. We may write the measured effect as being proportional to ($\mu E/k_B T$), where E is the applied electric field, T is the sample temperature, μ is the molecule's dipole moment and k_B is the Boltzmann constant; both (μE) and ($k_B T$) have units of energy. The Kerr effect is a means of generating complexity or order out of the chaos of the thermally driven motion in a sample of randomly oriented molecules (a gas, a liquid or a disordered glass).

The Kerr effect is an example of induced birefringence, the generation of a material which is capable of supporting two routes for the propagation of light beams. The best example in nature of such birefringence is calcite. Whereas in the Kerr effect, the applied electric field generates a partial, temperature-dependent ordering of the molecules, in calcite the ions in the crystal structure are perfectly regimented or ordered within the lattice. In particular, it is the optically anisotropic (planar) carbonate anions (CO_3^{2-}) that give to the crystal a birefringence of about 3° (that is, the difference between the parallel and perpendicular components of the refractive index are separated by 3°, or $n_{parallel} - n_{perpendicular} \sim 0.1$ radian), and allows the crystal to demonstrate a large birefringence.

If one views a pencil line through a crystal of calcite, the line appears twice because light can take two paths through the crystal (parallel or perpendicular to the oriented flat carbonate anions in the solid structure). The natural birefringence of the calcite crystal is huge and demonstrates the degree of internal structure and order of the crystal. The Kerr effect induced by application of a laboratory electric field to a sample of gas is much more modest. Whereas the natural ordering of carbonate anions in calcite generates a birefringence of 3°, the largest laboratory electric fields generate a birefringence or imposed order in gas samples of about 10^{-5} degrees. And that imposed order exists only as long as the electric field is applied; the crystal on the other hand can be chipped from a mountain and is naturally ordered.

In experiments involving the Kerr effect, the degree of ordering induced in the sample of material by the applied electric field is measured, and this allows the determination of the magnitude of the molecular dipole moment. Molecules that have a large electric dipole moment (for example, nitrobenzene and liquid crystals) have very large, easily measureable Kerr effects. In these experiments, if the measurement is made at a number of temperatures one can separate the temperature-dependent term that contribute to the measurement from the temperature-independent term. The temperature-independent term arises because the applied field distorts the molecules as well as orienting them.

Indeed, in experiments such as the Kerr effect or the quadrupole measurement, which we will consider below, one is ordering the molecules of interest by application of a uniform electric field or by application of a field gradient, respectively, and then one probes the degree of inducted orientation. This is possible because the molecules under study are anisotropically polarizable; that is, there is a difference in the degree to which the molecule is susceptible to an optical electric field (the laser beam) applied parallel or perpendicular to the main rotational axis of the molecule.

8.3 Experimental details

We will now look at the details of this experiment; in particular, at those elements that allow us to estimate the limits of the precision of the experiment for measuring electric quadrupole moments. The measurement of molecular quadrupole moments with the apparatus described here is typical of a direct measurement; that is, an experimenter measures an output parameter, and is able to calculate from this single quantity the property of interest. To extract the information of interest, there is no need for the measured data to be fitted to a computer model.

Although the experiment to measure quadrupole moments is relatively straight-forward, in practical terms it is in fact quite complex. It can be thought of as several 'black boxes', each of which is an experiment in its own right, and each of these individual experiments must be working optimally, and they must all be working coherently to allow a measurement of a molecular quadrupole moment. The experimenter induces an effect by applying an electric-field gradient to a sample of gas, and by applying an out-of-phase birefringence signal from a calibrated source (a nulling Kerr effect cell) in the optical train after the quadrupole cell; thus measuring directly the induced effect, which is the product of the applied electric-field gradient (a known quantity) and the molecular quadrupole moment (the unknown quantity). The output signal is monitored and after applying known constants, you have a direct measurement of the property of interest; that is, there is no lengthy computer-based data fitting. All this is a matter of care and patience on the part of the experimenter, and a different experimenter might be more careful or be less patient, characteristics that will contribute to the Type B measurement uncertainty (see section 5.3).

The experimental set-up is shown schematically in figure 8.1. The light from a He−Ne laser (red light at 632.8 nm) is polarized by passage through a Glan−Thompson calcite polarizer. This linearly-polarized light then propagates the length of the 1 m quadrupole cell containing the gas or vapour subject to a stable electric-field gradient (this is generated by two wires at the same high electrical potential on either side of the laser beam inside the Earthed quadrupole cell. At the centre of the laser beam, the electric field arising from the charged wires is zero, but the field gradient is a maximum.) As this linearly-polarized light propagates through the molecules oriented by the applied electric-field gradient, it becomes elliptically-polarized, as in the Kerr effect (see images (i) and (ii) in part (b) of figure 8.1). After the quadrupole cell, there is a quarter-wave plate ($\lambda/4$), which is used to facilitate measurement, and a calibrated Kerr cell that generates an out-of-phase Kerr effect in a calibrated (liquid CS_2) Kerr cell, which is used to null the ellipticity generated in the quadrupole cell. After this Kerr cell, there is the analyzing Glan−Thompson calcite polarizer and the photomultiplier detector.

With the laboratory z axis along the light path down the axis of the quadrupole cell, the x axis is chosen such that relative to it the azimuth of the linearly-polarized laser beam emerging from the first polarizer is precisely $\pi/4$. The azimuth of this linear polarization is chosen as the reference from which the azimuths of the other optical elements are measured. The quadrupole cell is then rotated about its axis until the plane of its two wires coincides with the yz plane.

The electric field of the light beam entering the quadrupole cell may be resolved into components ε_x and ε_y, which experience refractive indices n_x and n_y, respectively, as the beam propagates through the metre-long cell. If the path-length of the beam in the applied field gradient is l, the two field components will emerge

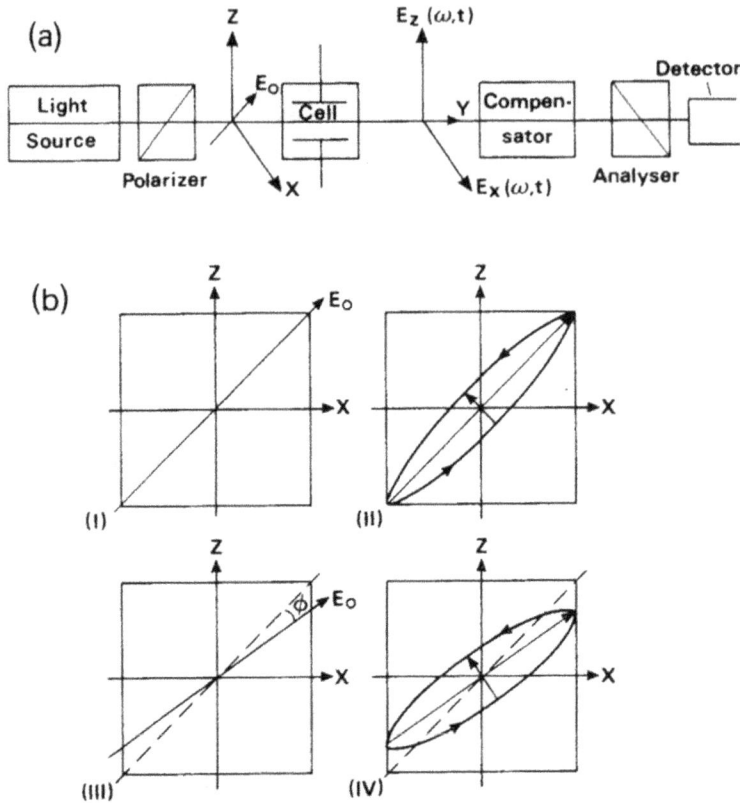

Figure 8.1. The basic components of an apparatus to measure an induced birefringence are shown in the upper part of this image. The birefringence is induced in a fluid in the cell, by application of an electric field, an electric-field gradient or by an applied magnetic field. The light source is a laser, the polarizer and analyzer are high-quality Glan–Thomson prisms. The lower part of this figure shows the polarization states in such electro-optic experiments: (i) linearly-polarized light propagating away from the observer with its electric vector at 45° or 135° to the x and z axes; (ii) elliptically-polarized light generated by the positive value of ($n_{parallel} - n_{perpendicular}$), the induced birefringence; (iii) positive rotation, of magnitude φ, of the plane of polarization produced by a positive value of ($k_{parallel} - k_{perpendicular}$), an induced dichroism; (iv) elliptically-polarized light, with its major axis rotated, produced by positive values of both ($n_{parallel} - n_{perpendicular}$) and ($k_{parallel} - k_{perpendicular}$). Images taken from Williams J H 1993 Aspects of the optical Kerr effect and Cotton–Mouton effect of solutions *Advances in Chemical Physics: Modern Nonlinear Optics* Part 2 vol 85 ed M Evans and S Kielich (Hoboken, NJ: Wiley).

from the quadrupole cell with a relative phase-difference δ given by (see images (i) and (ii) in part (*b*) of figure 8.1):

$$\delta = 2\pi l(n_x - n_y)/\lambda. \tag{8.1}$$

The light beam exiting the cell containing the oriented quadrupolar molecules is elliptically-polarized. But after passage through the quarter-wave plate, with its fast axis set at an azimuth of $\pi/4$, the beam emerges in a state of linear polarization with an azimuth of $\pi/4 + \delta/2$. The consequence of the combination of the quadrupole cell and the quarter-wave plate is to rotate the plane of polarization of the laser beam by $\delta/2$ (as shown in image (iii) in part (b) of figure 8.1).

Through choice of the length of the quadrupole cell, the magnitude of the applied electric-field gradient, and the pressure of gas in the cell, the apparatus was designed to measure induced retardations, δ, of order, 10^{-7} radian (this is the precision of the system, and the level of uncertainty arising from the system many be seen by looking at the data given in figure 4.1 in section 4.1). This means that we are seeking to measure a field-induced difference in the refractive index of the gas, $(n_x - n_y)$, of order, 10^{-16} (see equation (8.1)). By using a blue laser instead of the red laser, we see from equation (8.1) that we could increase the sensitivity of this measurement, as there would be more waves per centimetre.

There are very few phenomena in nature that can be measured to such a precision (the definitions of the second and the metre are the only others that come to mind). However, the magnitude of δ, the retardation induced by the applied electric-field gradient, means that it cannot be measured by extinction at the analyzing polarizer (but one can see that its size is linear in the length of the cell; so the longer the cell, the bigger will be the measured effect. Likewise, the shorter the wavelength of the light used for the measurements, the bigger the effect. So, the best design for this experiment would be a long path-length for the quadrupole cell (size limited by the practicalities of controlling its temperature uniformly) and a blue laser; δ is typically, of order, 10^{-6} radian, however, modulating the applied electric-field gradient and using a phase-sensitive detector to discriminate the signal of interest from the general (un-modulated) noise allows one to measure the small induced retardation in the laser beam arising from the oriented molecules in the 1 m quadrupole cell.

For a sample of molecules such as carbon dioxide, the measured induced birefringence, $n_x - n_y$ is related to the properties of the individual molecules through the expression (first derived by Buckingham)

$$n_x - n_y = \left\{ NE'/15\varepsilon_0 \right\} \left\{ (15\,B/2) + (\Theta\Delta\alpha/k_BT) \right\} \tag{8.2}$$

where N is the number density of molecules, E' the applied electric-field gradient, T the sample temperature and k_B is the Boltzmann constant, and ε_0 is the permittivity of free space (the degree to which an electric field may propagate). The first term in the second set of curly brackets gives the temperature-independent distortion of the molecule by the applied electric-field gradient, and the temperature-dependent term, $\Theta\Delta\alpha$, consists of the molecular quadrupole moment Θ and the anisotropic part of the polarizability of the molecules, $\Delta\alpha$. If the molecules have an axis of rotation C_n where $n \geqslant 3$, then $\Delta\alpha$ can be written as $\alpha_{parallel} - \alpha_{perpendicular}$; that is, the difference of the molecular polarizability parallel and perpendicular to the main rotational axis of the molecule.

We see from equation (8.1) and equation (8.2) how a series of measurements of the inducted birefringence, as a function of sample temperature will allow the separation and measurement of the two terms in the second set of curly brackets (this is the data displayed in figure 4.1 in section 4.1). However, to determine the sign and magnitude of Θ, one must know the sign and magnitude of $\Delta\alpha$; but the anisotropy of the molecule's polarizability is available from light scattering experiments (measuring the depolarization ratio). However, care must be taken as the errors in the measurement of $\Delta\alpha$ will carry through to the determination of Θ.

What has been outlined above is an ideal experiment, the perfect experiment. But experiments are performed in the real world, and are always imperfect. For example, the polarizers are almost crossed in the experiment so as to limit the total light level at the detector; one is looking for a small signal against a large background of noise, so any effect that increases stray light levels at the detector is to be avoided. On this point, we must consider the optical properties of the windows of the quadrupole cell and the Kerr cell to the induced birefringence. The strain birefringence induced by the window will not be modulated at the frequency of the applied field gradient in the quadrupole well, which is also the frequency of modulation of the nulling Kerr effect, but it will allow extra light to reach the detector, and so contribute to the final signal to noise problem of the measurement.

So the question becomes, is there a means of investigating in detail the contribution of the various optical components in the train of the experiment to the final measurement? That is, how do you quantify the contributions of the various optical elements to the measurement so as to maximize the measurement capacity of the experiment, by reducing the measurement uncertainty (that is, to increase the precision of the apparatus).

To determine the influence of each component on the polarization state of the light beam we can use Jones calculus. Piazza *et al* [3] have shown that the Jones matrices for certain optical components may be expressed as linear combinations of the unit and Pauli matrices

$$\mathbf{I} = \begin{pmatrix} 1 & 0 \\ 0 & 1 \end{pmatrix}, \qquad \mathbf{i} = \begin{pmatrix} i & 0 \\ 0 & -i \end{pmatrix}, \qquad \mathbf{j} = \begin{pmatrix} 0 & 1 \\ -1 & 0 \end{pmatrix}, \qquad \mathbf{k} = \begin{pmatrix} 0 & i \\ i & 0 \end{pmatrix}, \quad (8.3)$$

which combine in the following ways:

$$\mathbf{i}^2 = \mathbf{j}^2 = \mathbf{k}^2 = -1$$

$$\mathbf{ij} = \mathbf{k}, \qquad \mathbf{jk} = \mathbf{i}, \qquad \mathbf{ki} = \mathbf{j}. \qquad (8.4)$$

Then the Jones matrix for a linear retarder (phase retarders introduce a phase shift between the vertical and horizontal component of the field and thus change the polarization of the beam) of retardance ρ and azimuth ϕ is given by

$$\mathbf{J}(\rho, \phi) = \cos{(\rho/2)}\mathbf{I} + \sin{(\rho/2)}\cos{2\phi}\mathbf{i} + \sin{(\rho/2)}\sin{2\phi}\mathbf{k}, \qquad (8.5)$$

and that for an optical rotator having a rotation ψ is given by

$$\mathbf{R}(\psi) = \cos{\psi}\mathbf{I} + \sin{\psi}\mathbf{j}. \qquad (8.6)$$

The Jones matrix for a polarizer of azimuth σ has the form

$$\mathbf{P}(\sigma) = \begin{pmatrix} \dfrac{1}{2}(1 + \cos 2\sigma) & \cos \sigma \sin \sigma \\ \cos \sigma \sin \sigma & \dfrac{1}{2}(1 - \cos 2\sigma) \end{pmatrix} \tag{8.7}$$

but which cannot be expressed in terms of the $\mathbf{I}, \mathbf{i}, \mathbf{j}$ and $\mathbf{k}$ matrices. The normalized Jones vector for a linearly polarized light beam of azimuth η is

$$\nu(\eta) = \begin{pmatrix} \cos \eta \\ \sin \eta \end{pmatrix}. \tag{8.8}$$

Suppose that the polarization azimuth of the light entering the quadrupole cell is $\pi/4$, and that the analyzing polarizer is offset from its exact crossed position by a small angle α to simulate addition light reaching the detector against which one is attempting to make a measurement. Then the Jones vector for the light leaving the analyzer is given by

$$\nu = \mathbf{P}\left(-\frac{\pi}{4} + \alpha\right) \mathbf{M}_n \mathbf{M}_{n-1} \dots \mathbf{M}_1 \nu_0, \tag{8.9}$$

where $\mathbf{M}_1, \mathbf{M}_2 \dots \mathbf{M}_n$ are the Jones matrices for the retarders and the rotators in the optical train. Since the $\mathbf{I}, \mathbf{i}, \mathbf{j}, \mathbf{k}$ matrices form a closed group under matrix multiplication, it follows that the the product $\mathbf{M}_n \mathbf{M}_{n-1} \dots \mathbf{M}_1$ may also be expressed as a linear combination of these matrices,

$$\mathbf{M}_n \mathbf{M}_{n-1} \dots \mathbf{M}_1 = a\mathbf{I} + b\mathbf{i} + c\mathbf{j} + d\mathbf{k}. \tag{8.10}$$

If I_0 is the intensity of the light beam after passing through the first polarizer, then the intensity of the light beam at the detecting photomultiplier tube is given by

$$I = \nu^* \, \nu I_0,$$

which together with equations (8.7)–(8.10) yields

$$(I/I_0) = b^2 + c^2 + 2\alpha(ac + bd) + \alpha^2(a^2 - b^2 - c^2 + d^2),$$

in which we have retained terms to order α^2. Thus, to answer our question about how to maximize the measurement signal, we now have to find the various coefficients a, b, c and d for a given optical train.

The optical train of this experiment to measure molecular quadrupole moments is given in figure 8.1, we will now consider the addition of two additional optical components, the two windows of the quadrupole cell (these windows are typically of very dense glass and about 5 mm in thickness) and how they contribute to the measurement.

Consider the entry window of the quadrupole cell as a linear retarder of small retardance β_1 and arbitrary azimuth θ_1:

$$\mathbf{S}_1(\beta_1, \theta_1) = \cos(\beta_1/2)\, \mathbf{I} + \sin(\beta_1/2) \cos 2\theta_1 \, \mathbf{i} + \sin(\beta_1/2) \sin 2\theta_1 \, \mathbf{k}.$$

A similar expression could be written for the exit window of the quadrupole cell. Ideally, the quadrupole cell behaves as a linear retarder with retardance δ and zero azimuth. However, the cell was rotated into its experimental position, so one could allow for a small residual error in the azimuth, γ. The Jones matrix for the quadrupole cell is then:

$$\mathbf{J}q(\delta, \gamma) = \cos(\delta/2)\,\mathbf{I} + \sin(\delta/2)\cos 2\gamma\,\mathbf{i} + \sin(\delta/2)\sin 2\gamma\,\mathbf{k}.$$

So, the possible contributions of the extra, but essential optical elements in the optical train to the final measurement may be isolated and examined.

In this way, the Jones matrices for every component in the optical train may be combined using equation (8.10), and an expression for the final intensity of light reaching the detector be generated. In practice this is a long and tedious calculation, because of the many unknown quantities that appear. Making only small angle approximations and retaining terms to a certain order does simplify the algebra, but there remain many unknowns.

$$(I/I_0) = \delta\left\{ C_1 - C_2 + \varepsilon\left[\mp\cos\phi + 2S_2(1 + \sin\phi)\right] + \alpha(\pm\cos\phi - 2S_2\sin\phi)\right\}$$
$$+ \theta\left[-2(C_1S_1 + C_2S_2 + 2C_1S_2)\sin\phi \pm 2(C_1 + C_2)\cos\phi - 2\varepsilon(1 + \sin\phi) + 2\alpha\right].$$

Here

$$S_n = (\beta_n/2)\sin 2\theta_n, \quad \text{and} \quad C_n = (\beta_n/2)\cos 2\theta_n,$$

and ε and ϕ arises from the manner in which one models the quarter-wave plate. The value of θ equal to $\delta/2$ is termed θ_{null}, as it is this rotation which, for ideal optical components in perfect orientation, is required to null the birefringence generated in the quadrupole cell. This induced retardation may be determined by writing a computer programme to sample the signal coming from the phase-sensitive detector as a voltage is applied to the nulling, calibrated Kerr cell.[2] From the above equations it follows that, for a small value of ϕ, these lines will intersect at a value of θ, θ_{int}, given by

$$\theta_{int} = -(1/2)\delta(\pm 1 - 2S_2/1 + \phi).$$

Such an analysis allows one to determine the magnitude of the contribution of the windows to the measured effect, and if a systematic study of the contributions of all of the windows were made, at differing orientations relative to the direction of polarization of the incident light beam (the windows might for example possess an intrinsic strain, which could induce an ellipticity in linearly-polarized light), one may seek to minimize the contribution of the widow to the measured effect. With such a calculus, one could also investigate the most appropriate composition of the windows (quartz, hard glass or soft glass, etc) to maximize the precision of the apparatus. However, pressure resistant windows are essential.[3]

---

[2] Graham *et al* [4] have performed the most detailed analysis of the contributions of the various optical elements in this experiment to the overall noise of the experiment. For example, they found θ_{null} from the intersection of two graphs of I/I_0 versus θ corresponding to two different offsets of the quarter-wave plate.

[3] In the late-1970s when I was using such an apparatus, there were no laptop computers, and the data coming from the lock-in detector went to a buffer and was then punched onto paper tape, which was then read into the university mainframe for subsequent analysis.

Both the quadrupole moment determination and the measurement of the Kerr effect (and the optical Kerr effect that we will encounter in chapter 9) are experiments that are undertaken between crossed polarizers. In such situations, the induced retardation δ arising from: the electric field in the Kerr effect, the electric-field gradient in the quadrupole effect, the applied optical electric field in the optical Kerr effect, or an applied magnetic field as in the Cotton–Mouton effect, is determined by placing the birefringent element between crossed polarizers (see figure 8.1) with its birefringent axis oriented at 45° to the polarization direction. The intensity of the light transmitted by the analyzing polarizer is, for small δ

$$I = I_0 \sin^2 (\delta/2) \sim I_0(\delta/4).$$

where I_0 is the incident light intensity; typically, δ is of order 10^{-6} radian and, because of the presence of stray birefringence of order 10^{-4} radian, which may be quantified and identified using the Jones calculus described above, a heterodyne method (modulation of the applied field and phase-sensitive detection of the output) is usually adopted for measuring the small inducted retardation. If both a small modulated birefringence $\delta(\omega)\sin\omega t$, and a large static (dc) birefringence δ_0 are placed between the crossed polarizers, then at the detector

$$I = I_0 \left[\sin^2 (\delta_0/2) + (1/2) \sin \delta_0 \delta(\omega) \sin \omega t + \text{order}\big(\delta(\omega)\big)^2 \right].$$

A phase-sensitive detector measures the component of I oscillating at the frequency ω. This term may be made larger by increasing the magnitude of δ_0. However, the noise level against which you are seeking to measure $\delta(\omega)$ will also become larger as it will be determined by the total light intensity passing the analyzing polarizer, and hence will be strongly dependent on δ_0^2. As mentioned above, in practice, δ_0 is introduced by a linear retarder and may be, of order, 0.1 radian.

At the detector (a red-sensitive photomultiplier tube), the average number of photons counted in a time interval Δt is given by $N = IQ\Delta t$, where Q is the quantum efficiency of the detector. The number of photons detected in this time interval will be randomly distributed about a mean with a standard deviation $\Delta N = \sqrt{N}$. These statistical fluctuations are called shot noise and determine the sensitivity of the detection system for a particular measurement. For a typical induced birefringence apparatus, $\delta(\omega) \ll \delta_0$, and the signal to noise ratio is:

$$\frac{S}{N} = (1/2)QI_0\Delta t \sin \delta_0 \delta(\omega)/\sin (\delta_0/2) \ \sqrt{QI_0\Delta t} = \sqrt{QI_0\Delta t} \cos (\delta_0/2)\delta(\omega).$$

Hence, δ_0 should be chosen large in relation to the stray birefringence but small compared to 1 radian; the signal to noise ratio is then independent of δ_0. For example, a He–Ne laser emitting about 10^{16} photons per second at 632.8 nm, assuming $Q = 0.1$ and $\Delta t = 1$ second, we have $\delta(\omega) = 3 \times 10^{-8}$ radian as the induced retardation giving a signal/noise of unity. But we see from equation (8.1) that this sensitivity could be increased by using a longer quadrupole cell and by using a blue laser (which would increase the sensitivity in the ratio 632.8 nm/458 nm).

We have looked in some detail at how one would make a measurement of the molecular quadrupole moment of a molecule. This is what I term a direct measurement. Here one constructs an apparatus, it can be quite a complex apparatus, but one measures only one thing. In the quadrupole moment experiment, one is only interested in the magnitude and sign of the signal coming from the phase-sensitive detector, which is analyzing the signal coming from the photomultiplier tube, which is measuring the amount of light passing the analyzing polarizer. There is only one measurement of interest. As mentioned, the experiment can be repeated at a number of sample temperatures to separate the temperature-dependent and the temperature-independent terms that contribute to the induced birefringence, and which both represent information about the electronic structure of the molecule under investigation (see figure 4.1 in section 4.1). But even if a series of temperature measurements are made, each one will only be a measurement of the one single observable, the voltage coming from the phase-sensitive detector.

The observer measures one number, the retardation arising from the applied electric-field gradient, and then multiplies it by a number of other parameters (see equation (8.2)), the sample temperature, the number density of molecules in the quadrupole cell, the polarizability anisotropy of the molecule, the value of the electric-field gradient arising from the potential applied to the two wires inside the Earthed metal cylinder, all of which will possess an intrinsic measurement uncertainty. Then he or she has a value for the quadrupole moment of the molecule under study. It is a classic experiment; you measure one observable to determine one unknown. We have seen that Jones calculus may be used to analyze the contributions of the various optical components to the noise in the experiment, and how one is thus able to modify or discriminate against such sources of unwanted noise. After that, one attempts to make each component in the apparatus as reliable and precise as possible, and one then makes measurements, seeking reliability in a series of measurements, which may then be subject to some statistical analysis.

This final statistical analysis need not be complex and laborious, because analysis (as outlined above) has shown that the apparatus is not designed to readily measure induced retardations, of order, 10^{-8} radian or smaller. One may make such measurements, but you know from your analysis of the apparatus that such measurements will be subject to large amounts of random noise, and so will never be very reproducible and will, consequently, be very uncertain, and so do not warrant a great investment of time.

8.4 How many measurements do you need?

Before one gets into the details of how to analyze the data one has measured, it is as well to look briefly at how many measurements are needed.

8.4.1 Over-determined system

In mathematics, a system of equations is considered overdetermined if there are more equations than there are unknowns. An over-determined system is almost always inconsistent (that is, it has no solution) when constructed with random coefficients. However, an over-determined system will have solutions under some circumstances; for example, if some equation occurs several times in the system, or if some equations are linear combinations of the other equations. For the experimenter, this set of abstract equations becomes the set of equations that define what it is that one is measuring (one of the unknown coefficients) and the known experimental conditions and parameters (the known coefficients).

Each unknown in the set of equations can be thought of as an available degree of freedom. Each equation introduced into the ensemble can be viewed as a constraint that restricts one degree of freedom. However, for every variable giving a degree of freedom, there exists a corresponding constraint. The over-determined case occurs when the system has been over-constrained; that is, when the equations outnumber the unknowns. In contrast, the under-determined case occurs when the system has been under-constrained; that is, when the number of equations is smaller than the number of unknowns.

Consider the system of three equations and two unknowns (x_1 and x_2), which is overdetermined: $2x_1 + x_2 = -1$, $-3x_1 + x_2 = -2$, $-x_1 + x_2 = 1$. There is one solution for each pair of linear equations: for the first and second equations (0.2, −1.4), for the first and third (−2/3, 1/3), and for the second and third (1.5, 2.5). However there is no solution that satisfies all three simultaneously. If one were to graph these equations, one would observe configurations that are inconsistent, because no point is on all three graphed lines.

The only cases where the over-determined system does in fact have a solution are: one equation is linearly dependent on the others and there is an intersection, the three lines intersect at the same point, and when the three lines are superimposed (an infinity of intersections). These exceptions can occur only when the over-determined system contains enough linearly dependent equations that the number of independent equations does not exceed the number of unknowns. Here, linear dependence means that some equations can be obtained from linearly combining other equations. For example, $y = x + 1$ and $2y = 2x + 2$ are linearly dependent equations because of the constant of multiplication. There is no, one measurement that will satisfy the three equations and yield all the unknowns. A series of experiments is required. In the quadrupole moment experiment, one could write down a set of equations governing the various 'black box' components of the overall experiment, but these equations would to a large degree be dependent upon each other.

8.4.2 Under-determined system

In mathematics, a system of linear equations or a system of polynomial equations is considered under-determined if there are fewer equations than unknowns. Each unknown can be seen as an available degree of freedom. Each equation introduced into the system can be viewed as a constraint that restricts one degree of freedom. Therefore, the critical case (between over-determined and under-determined) occurs when the number of equations and the number of free variables are equal. For every variable giving a degree of freedom, there exists a corresponding constraint removing a degree of freedom. The under-determined case, by contrast, occurs when the system has been under-constrained; that is, when the unknowns outnumber the equations.

An under-determined linear system has either no solution or infinitely many solutions; for example, $x + y + z = 1$ and $x + y + z = 0$ is an under-determined system without any solution; any system of equations having no solution is said to be inconsistent. On the other hand, the system $x + y + z = 1$ and $x + y + 2z = 3$ is consistent and has many solutions. All of these solutions can be characterized by first subtracting the first equation from the second, to show that all solutions obey $z = 2$; using this in either equation shows that any value of y is possible, with $x = -1-y$.

There are algorithms available to decide whether an under-determined system has solutions, and if it has any, to express all solutions as linear functions of the variables; the simplest would be Gaussian elimination.

But back to the experimental laboratory. It is not possible to solve a problem where the number of unknowns exceeds the number of equations describing the system of interest. If you have two unknowns, which are to be determined experimentally, but you have only one equation describing the observable phenomenon, you cannot measure the two unknowns independently. The best you can do is determine a ratio of these two unknowns. If then, this ratio is used as a known quantity in some other experiment, then one needs to be careful. Ratio measurements can only be used to attribute absolute values to, for example, unknown measurement standards if, at least, one absolute reference value is known at the starting point of the ratio chain. In addition, measurement results cannot be more accurate that the standard used in the measurement process.

References

[1] Buckingham A D 1959 *J. Chem. Phys.* **30** 1580
[2] Debye P 1963 *quoted in Chem. Eng. News* **41** 40–3
[3] Piazza R, Degiorgio V and Bellini T 1986 *Opt. Commun.* **58** 400
[4] Graham C, Pierrus J and Raab R E 1989 *Mol. Phys.* **67** 939

Chapter 9

Indirect measurement: the optical Kerr effect

9.1 Introduction

In the previous chapter we looked at what I term a direct measurement; that is, one where you build an apparatus to measure one observable. This observable is then multiplied by constants and fixed or defined parameters, and you derive the quantity of interest. The error analysis of such experiments is quite straightforward as one can analyze the possible sources of noise to the one observable of interest and determine the ultimate sensitivity of the apparatus. But what happens when there is no straightforward observable to the experiment; for example, when one is making an indirect measurement of something?

By indirect measurement, I am referring to an experiment where the measured data must first be computer fitted to some theoretical model before any discussion of the measured result can be entertained; that is, experimental data fitting. Today, this is how a great many of the most significant discoveries in physics are made. It is not a scientist working in a laboratory with an apparatus he or she has built. Today, discoveries such as the Higgs boson or the observation of gravitational waves involve huge teams of scientists measuring and recording vast amounts of data, but where only a tiny part of that data contains the desired phenomenon. This may well be a phenomenon, which to be identified and quantified within the measured data requires a complex process of data fitting and data manipulation. For example, how was it determined that quarks are the ultimate basis of matter? The atom was known philosophically to the Ancient World, but was not identified until the late-19th Century, and not observed directly until the second-half of the last Century. The electron was identified in Cambridge at the end of the 19th Century by the group of Joseph John Thomson; the proton was discovered by Ernest Rutherford and his team at the end of World War I, and the neutron was discovered by James Chadwick in 1932. But what of the constituents of the proton and the neutron; how were they 'observed'?

doi:10.1088/978-1-6817-4433-9ch9

Well, the truth is that the constituent particles of the proton and the neutron (quarks) have never been 'observed' in the same way that the electron or the proton were originally observed and studied. Quarks have been studied by fitting the data that is accumulated by smashing protons together at high energy (the proton is electrically charged and so it may be accelerated electro-magnetically). Such experiments are performed at laboratories like CERN, the European organization for nuclear research, in Switzerland, and they generate vast quantities of data.

Approximately 600 million times per second, sub-atomic particles collide within the Large Hadron Collider (LHC) at CERN. Each collision generates particles that decay in complex ways into other sub-atomic particles via well-established routes that may be analyzed using quantum mechanics and the Standard Model of particle physics. Circuits record the passage of each sub-atomic particle through a detector as a series of electronic signals, and send the data to the CERN computers for digital reconstruction. The digitized summary is recorded as a 'collision event'. Then physicists must sift through the 30 petabytes $(30 \times 10^{15}$ bytes) or so of data produced annually to determine if the collisions have (literally) thrown out any interesting physics or new particles.

When a piece of interesting physics or a suspected new particle is identified, an army of scientists home in on that particular event and analyze all the relevant computer records. This is not a direct experiment that was designed and undertaken, it is an event that has occurred, it just happened in the collider; it was one event in 6×10^8 events per second taking place in the LHC. The intense scrutiny of the physicists might, of course, come to nothing; one was merely seeing a well-known phenomenon from an odd angle (see figure 4.4), but sometimes new physics is observed.

For the 'observation' of the quark, protons were collided at higher and higher energies so as to probe their innermost constituents. The scattering event for the collision of two protons was observed, as was the expulsion of all the various particles in the two protons; all moving outwards from the scattering centre. Physicists start with a model of what should be happening according to their current best theory, and then look for evidence of that theorized event. With the quark, it was theorized that each proton is formed of three quarks. So when two protons collide at sufficiently high energy, the scattering event should contain evidence of the interaction of nine quarks. This was what was observed in the vast amount of accumulated data. Fitting the observed data to a model of nine interacting quarks explained what was being measured in the detectors of the atom smashers. And on going to even higher energies, no new physics or unexplained events were seen in the fitted data originating from the scattering event, so it could be assumed that the model of a proton consisting of three quarks, which are indivisible, is correct. In no case was there a direct 'observation' of a quark. There was an indirect observation; one starts with a model of what you expect and you look for evidence of that model. If no evidence is found, then the model must be modified or abandoned.

This was how quarks were established as the basis of matter, and how the Higgs boson was observed. Indeed, fitting an observed event or phenomenon to a model was how gravitational waves were observed. There was no direct measurement of a Higgs boson or a gravitational wave, but the observation of a scattering event at CERN that could only be modelled (and thus explained) by assuming that the particle involved was a Higgs boson with some of the expected properties of the Higgs boson. And the signal (GW150914) seen at both LIGO gravitational wave detectors could only be explained by a model that assumed that it matched the predictions, from general relativity, for a gravitational wave emanating from the inward spiralling and merging of a pair of black holes of around 36 and 29 solar masses, and the subsequent 'ring-down' of the resulting single black hole.

So, in such indirect experiments, you begin the experiment having a complex model of what may be happening (or of what is about to happen). Of course, the person who designed the quadrupole experiment discussed in the previous chapter also had in mind a theory of what happens when linearly-polarized light passes through a birefringent medium, and this model would have been subject to evolution should the experiment have yielded unexpected, but reproducible results. That is the nature of experimentation; a theory is useful, but the results of a good experiment will last forever.

In the indirect experiment, the model is all important. It is not possible to do a 'quick and dirty' measurement to check that there is something there to be measured, subsequently, with greater care and precision. In some way, indirect experiments are unique experiments—they take a long time to do. We have been looking for gravitational waves and the Higgs boson for half a century, and I am sure that now these things have been observed, there will be more observations in the near future. But these experiments are not susceptible to the type of statistical analysis that may be made on the measurements derived from a direct experiment, such as the determination of the quadrupole moment of a molecule. My own PhD thesis was based on measurements of molecular quadrupole moments. It required a couple of years to build that apparatus, and many hundreds of measurements were then made on more than a dozen gases and vapours. During this time, a trivial statistical analysis established the precision of my measurements. But how do you establish error bars for the observation of a Higgs boson or of a gravitational wave?

9.2 The optical Kerr effect

As an example of an indirect experiment (that is, a measurement where one does not make a direct determination of the desired quantity, but must analyse the measured data via a process of data fitting), we will consider the optical Kerr effect (OKE) generated by pulsed laser systems.

As we saw earlier, the Kerr effect was traditionally used to make measurements of static molecular properties; in particular, the bulk electric susceptibility of a fluid. The laboratory electric field that is applied to the fluid to generate the birefringence is a static field. This applied field orients the molecules, and when linearly-polarized light passes through these oriented molecules it becomes elliptically-polarized. Measurement of this induced ellipticity allows investigation of the electrical properties (electric dipole moment and the anisotropic part of the molecular polarizability) of the molecules under investigation. The applied electric field may be modulated to assist in improving the signal to noise of the measurement (as described above), but this frequency of oscillation is so low (a few hundreds of cycles per second) that the molecules are able to follow the reversing field.

In the OKE, however, the applied electric field that induced the birefringence in the fluid oscillates at optical frequencies.[1] Consider a pulsed laser with a pulse width of about 30 nanoseconds with a peak power (W) of 40 MW cm^{-2}. We may calculate the peak electric field of the light as $E = \sqrt{W/c\varepsilon_0}$ (where c is the speed of light and ε_0 is the permittivity of free space) to be 1.2×10^7 V m^{-1}, which is larger than any available laboratory applied, static electric field. The associated magnetic field, which would be a source of high-frequency magnetic birefringence is a factor of c smaller, and so is negligible. Given the sensitivity of the various experiments designed to measure induced retardations of linearly-polarized light beams by oriented molecules (as show above), it is even possible to determine the OKE induced by a continuous laser of modest power. However, in the experiments to be discussed here, the laser inducing the orientation is a pulsed laser with a pulse width of 200 fs and a peak power for orienting the molecules of interest of 10^9 W, giving an electric field, of order, 10^7 V m^{-1} in a weakly focused (beam waist, of order, 150 μm) beam.[2]

The systems investigated below are aqueous solutions of simple ions. That is, systems that it is only possible to study with the OKE. If you take an aqueous solution of sodium chloride and apply a static or dc electric field, you will not orient the ions, but you will cause the ions to move in the solution, and you will have electrolysis. Pure water is also slightly conducting, but a static Kerr effect constant has been measured for pure water at 632.8 nm of $2.96 \pm 0.08 \times 10^{-14}$ V^{-2} m. If, however, the fluid is subjected to an applied electric field oscillating at 10^{15} s^{-1} (the optical field of a pulsed laser) the dipolar contribution to the induced birefringence is lost because the molecules cannot follow the rapidly oscillating electric field, and the OKE of pure water has been found to be 2.48×10^{-16} V^{-2} m (see [2]). What is clear is the loss of the dipolar contribution to the induced birefringence. In the static measurement, the molecular dipole moment dominates, but at optical frequencies, the measurement probes only the distortion of the anisotropic part of the molecular polarizability by the applied optical field.

A schematic representation of the apparatus used for the OKE measurements that will be discussed below is given in figure 9.1. The experiments to be discussed were carried out in the European Laboratory for Nonlinear Spectroscopy, Florence, Italy. The 80 ps pulses of a 12 W mode-locked Nd:YAG laser are pulse-compressed and frequency doubled to 4 ps at 532 nm. This beam was stabilized and used to pump a dye-laser, the output of which, 250 mW at 600 nm with a 350 fs width, is again compressed to give 120 fs pulses.

---

[1] This orienting electric field, is the field associated with the light beam.

[2] The optical Kerr effect was predicted by Buckingham in 1956 [1], well before the advent of lasers. On being asked by the author how he (ADB) had envisaged making a measurement of this phenomenon in 1956, the response given was that the light from a WWII searchlight could have been focused down into a suitably polarizable fluid.

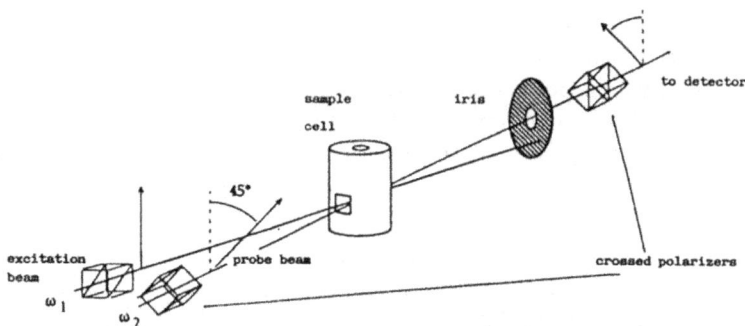

Figure 9.1. Schematic representation of a typical femtosecond optical Kerr effect laser system (the TROKE experimental set-up at the European Laboratory for Nonlinear Spectroscopy, Florence, Italy). The two incoming laser beams are at frequencies, ω_1 and ω_2; the former, the excitation beam, is the more intense and induces a birefringence in the medium under investigation, which is then probed by the weaker probe beam. The polarization of the probe beam is at 45° to that of the excitation beam and is crossed with respect to the analyzer or second polarizer. The iris defines the experimental geometry, and serves as a beam stop. Image taken from [2].

In the OKE experiments, the more powerful orienting laser pulse, which is linearly-polarized, generates an anisotropy in the refractive index of the fluid under investigation (as the applied laboratory electric field does in the static Kerr effect measurements discussed above). This induced optical anisotropy of the fluid generated by the intense optical field is then probed by a weaker, variably delayed, probe pulse that is also linearly-polarized, but with the plane of polarization at 45° to the direction of polarization of the intense orienting pulse. The birefringence induced by the applied electric field oscillating at optical frequencies is then determined by observing the modulation in the intensity of light passing the analyzing polarizer. This apparatus could readily measure an induced retardation, of order, 1×10^{-7} radian, which corresponds to a difference in the parallel and perpendicular components of the refractive index of ($n_{\text{parallel}} - n_{\text{perpendicular}}$) of about 10^{-13} at 600 nm for a 1 cm path length. The quadrupole moment experiment described in the last chapter has greater precision, in that we could measure an induced retardation, of order, 3×10^{-8} radian, but our path-length was 100 cm.

In simple organic liquids (for example, liquid benzene), investigated by the OKE before these measurements on aqueous solutions, it was seen that five contributions to the complex dynamics of the molecules in the liquid state could be identified:

    i. an instantaneous, purely electronic response of the polarizability (the distribution of electrons) of the molecule, arising from the molecule's hyperpolarizability;

    ii. an intermolecular vibrational contribution originating from the motion of the target molecule moving in the cage of nearest neighbour molecules; a motion decaying with the lifetime of the nearest-neighbour cage ≤200 fs;

    iii. an intramolecular vibrational contribution, decaying with the dephasing time of the molecular vibrations, typically a few hundreds of femtoseconds;

iv. a term arising from molecular interactions; here relaxation is governed by the local structure or relative orientations of the molecules in the liquid state, and which decays over a few hundreds of femtoseconds;

v. a rotational contribution, the timescale of which is determined by the macroscopic properties of the fluid (sample temperature, density, viscosity, the size of the molecules, and which could range from a picosecond to tens of nanoseconds.

Thus in terms of fitting any measured data for subsequent analysis, the standard model adopted for the interpretation of the OKE in simple organic liquids involved five parameters each with a weighting coefficient; that is, there are of order 10 unknowns. The system that we chose to investigate did not consist of organic liquids, but aqueous solutions of simple salts, which we considered as being less complex in their dynamics by virtue of the stronger (ionic) intermolecular forces involved.

The OKE measures the change in refractive index of the aqueous solution due to the interaction of the ions present in the medium (and the medium itself) with the optical electric field. The time-resolved OKE signal $S(\tau)$ at a delay time τ, is given by the double integral

$$S(\tau) = \int_{-\infty}^{\infty} I_{\text{probe}}(t - \tau) \sin^2 \left[\int_{-\infty}^{t} R(t - t')I_{\text{pump}}(t')dt' \right] dt \qquad (9.1)$$

where τ is the delay between the orienting pulse of the laser and the probing pulse of the laser, $R(t)$ is the response function for the system and I_{probe} and I_{pump} are the intensities of the probe and orienting pump beams, respectively. The second integral in equation (9.1) represents the phase difference between the parallel and perpendicular components of the probe beam, induced by the pump beam (the desired observable). Assuming that this phase shift is small we may write:

$$S(\tau) = \int_{-\infty}^{\infty} I_{\text{probe}}(t - \tau) \left[\int_{-\infty}^{t} R(t - t')I_{\text{pump}}(t')dt' \right]^2 dt. \qquad (9.2)$$

The time profiles for the laser pulses were seen to be well fitted by use of a bi-exponential function with a fast rising and slow decaying form:

$$\exp(\alpha_k t) \quad \text{for} \quad t < 0$$
$$\exp(-\gamma_k \alpha_k t) \quad \text{for} \quad t > 0$$

where $\gamma > 1$ is an asymmetry parameter and $k = 1$ for the pump beam and $k = 2$ for the probe beam, respectively. The typical model for the response function $R_s(t)$ is the sum of exponential, nuclear, or molecular decays plus an instantaneous or electronic response; that is,

$$R_s(t) = R_{\text{electronic}}(t) + R_{\text{molecular}}(t)$$
$$= n_2{}^e \delta(t) + \sum (n_2{}^i / \tau_i) \exp(-t/\tau_i) \qquad (9.3)$$

where τ_i is the relaxation time of the ith decay component and $\delta(t)$ is the delta-function. The other terms are as follows: n_2^e represents the instantaneous electronic component to the observed induced birefringence (component (i) from the list above), and n_2^i are the various nuclear or molecular contributions to the observed relaxation process (vibrational and rotational).

We see immediately the difference between a direct experiment, such as the measurement of a molecular quadrupole moment and the more complex indirect experiment such as a determination of the OKE of a liquid. In the direct experiment one constructs an apparatus to measure one observable; all other experimental conditions are specified in advance. The errors in the measurement of the induced retardation propagate through to the derived value of the quadrupole moment. Provided the errors in the determination of the experimental conditions are lower than the errors involved in the determination of the induced retardation, then the value of the retardation and the quadrupole moment will have much ,the same measurement error. If, however, there is a much greater uncertainty in; for example, the determination of the number density of molecules in the quadrupole cell, then this error will define the final uncertainty in the quadrupole moment.

In the OKE experiment of ionic solutions, however, one measures a huge quantity of data, but this huge amount of data is required because you have many more unknowns to be simultaneously determined (perhaps as many as five). The laser generates pulses of light, of order, 100 fm in length; there are a lot of these pulses in an experimental run. In both the OKE experiment and the direct quadrupole measuring experiment, although one is a static measurement and the other a dynamic measurement, we are able to measure tiny fluctuations of the refractive index of a fluid, but they are only able to tell us something about these weak effects because there are many molecules contributing to the measurement. In addition, we are measuring (repeatedly measuring in the OKE) for a long time. About 10^{24} molecules contributing to the determination of the molecular quadrupole moment, and in the OKE experiment we have about 10^{22} ions contributing to the measurement. Also, in both cases we are able to measure for a long period; in the quadrupole moment experiment we are using a continuous laser beam and so we can integrate the signal from the phase-sensitive detector by varying the time-constant of the circuitry, and in the pulsed laser OKE experiment, we can count for as many pulses as we think necessary.

Both direct and indirect measurements have their place in science, but the latter only really became possible with the advent of computers.

In figure 9.2(a) we see the OKE as determined in pure water and in three aqueous solutions of sodium chloride. As the delay time of the probing pulse can be varied with respect to the powerful pumping pulse, we are able to observe the relaxation of the oriented molecules and ions. That is, the pump laser-pulse orients the molecules via the near instantaneous electronic interaction of the molecules (water and solvated ions in these figures) with the electric field of the laser-pulse. Then we vary the delay time and observe the OKE signal decay as the molecules relax or lose their induced orientation. This relaxation process can then be used to investigate intermolecular interactions via a suitable model which must be fitted to the measured data.

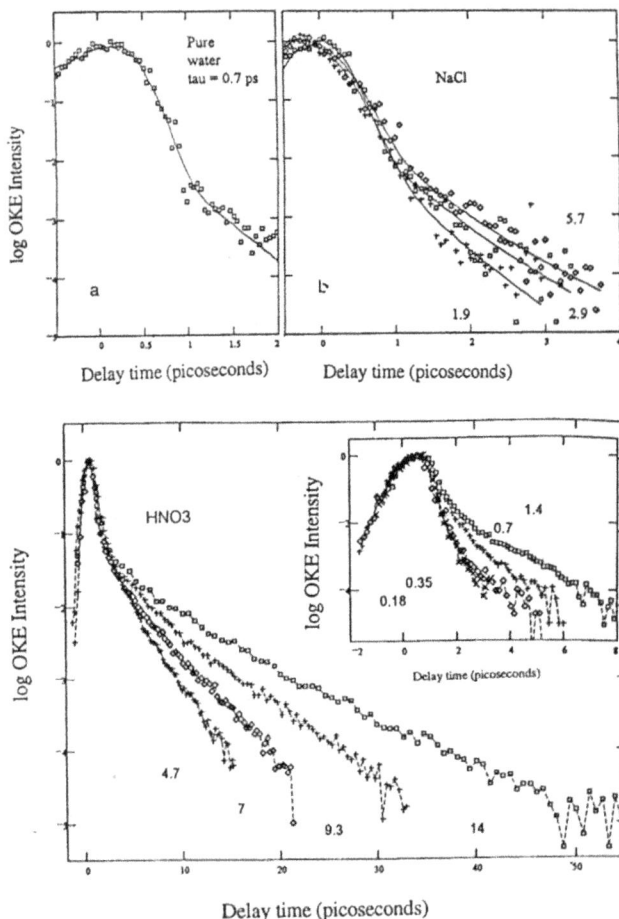

Figure 9.2. Typical relaxation data from femtosecond time-resolved optical Kerr effect measurements. First there is the intense electronic response of the sample to the applied optical electric field, then a slow fall of the signal, which contains information about the dynamics of the molecules and ions in the sample (an aqueous solution at room temperature). (*a*) The response of a sample of pure water and of three aqueous solutions of sodium chloride. (*b*) Relaxation data for eight solutions of nitric acid of varying concentration. As the nitrate anion is anisotropically polarizable, the signal (and hence signal to noise) from the nitrates is larger than that from the chlorides. The data is analyzed by assuming that the laser pulses (excitation and probe pulse) are δ-functions with infinitely sharp rise time; this is not actually the case and, consequently, there are some data before the time zero of the arrival of the probe pulse; this serves to demonstrate the uncertainty in the model. In this figure, the intensity axis has a log scale, and the time axis is in picoseconds (10^{-12} second). The data is taken from [2].

In pure water, the peak at a delay time of zero seconds is the electronic orientation of the molecules (the dipole moments of which cannot contribute). This is the distortion of the anisotropic part of the molecules' polarizability by the applied optical field. Then with time, this orientation decays away with a decay rate of 0.7 ps (for a sample of water at 18 °C). Similarly in the sodium chloride solutions, where in principle there should be no contribution from the sodium cations or the chloride anions because they are spherically symmetric. But one can clearly see the difference between the OKE of water and of the solutions, and this difference varies with solute concentration. The spread in the data points gives an idea of the measurement uncertainty; particularly in the solutions of sodium chloride, which should only generate a small OKE as there are no anisotropically polarizable ions present. However, we see in figure 9.2(b), the OKE generated in nitric acid of various concentrations. Here there is an anisotropically polarizable ion present (the planar NO_3^-) thereby generating a bigger signal than the solutions of isotropic ions. There is thus more signal, a better signal to noise and a lower measurement uncertainty. Figure 9.3 contains data for solutions of sodium nitrate (a), which generate a large OKE signal, and different concentrations of HCl (b), which generates a smaller OKE signal than that generated by the nitrate anions.

What is also apparent in these figures is the huge amount of data one is able to collect and subject to analysis. In the quadrupole moment experiment, one spent all day measuring one quantity half a dozen times, and then performing a trivial statistical analysis of the six measurements to determine the quadrupole moment. In the OKE experiment, in a couple of hours one can generate enough data to keep a student busy (with his or her computer fitting software) for days. What this greater quantity of data means is that one can seek to identify more than one unknown from the measured data—provided one has a suitable model.

We may write the Kerr constant for a multi-component system as

$$K_{\text{solution}} = \sum_i x_i K^{(i)} + \sum_{ij} x_i x_j K^{(ij)} + \cdots, \tag{9.4}$$

where x_i are the mole fractions of the components with $K^{(i)}$ being the corresponding Kerr constant. The first sum corresponds to the additivity of the signals arising from the various components present, and the second summation, and subsequent summations permit deviations from additivity. In the ionic solutions studied here, we can assume that at low concentrations all the sodium and chloride ions, or sodium and nitrate ions are completely solvated; that is, there is an octahedron of six water molecules around each Na^+ and Cl^- ion, and two water molecules above and below the planar nitrate ions. However, when the solutions become concentrated (6 moles per litre is very concentrated), there are barely enough water molecules to generate these solvation sheaths around all the ions. We have therefore a picture (a model) of long-range, polarizable structures established at low solute concentrations, which are disrupted by strong ionic forces produced by the addition of extra solute.

Combining the equations given above, our time-resolved OKE measurements were analyzed by means of the following equation (the model used to fit the data):

$$R_s(t) = x_{\text{water}}\{n_2^e(\text{water})\delta(t) + (n_2^o(i)/\tau(i))\exp-(t/\tau(i))\}$$

$$+ x_{\text{solute}}\{n_2^e(\text{solute})\delta(t) + (n_2^o(ii)/\tau(ii))\exp-(t/\tau(ii)) \tag{9.5}$$

$$+ (n_2^o(iii)/\tau(iii))\exp-(t/\tau(iii))\}$$

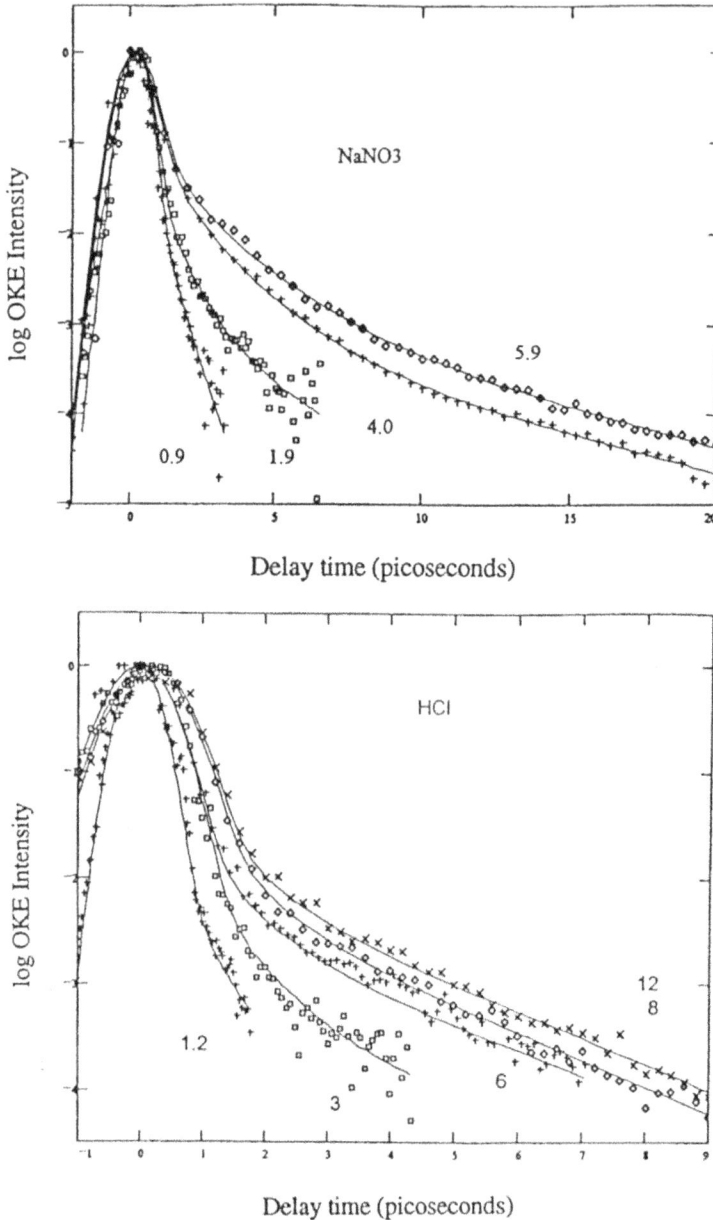

Figure 9.3. Typical relaxation data from femtosecond time-resolved optical Kerr effect measurements. First there is the intense electronic response of the sample to the applied optical field, then a slow fall of the signal, which contains information about the dynamics of the molecules and ions in the sample. (*a*) The response of four aqueous solutions of sodium nitrate. (*b*) Relaxation data for five solutions of hydrochloric acid of varying concentration. As the nitrate anion is anisotropically polarizable, the signal (and hence signal to noise) from the nitrates is larger than that from the chlorides. The data is analyzed by assuming that the laser pulses (excitation and probe pulse) are δ-functions with infinitely sharp rise time; this is not actually the case and consequently, there are some data before the time zero of the arrival of the probe pulse; this serves to demonstrate the uncertainty in the model. In this figure, the intensity axis has a log scale, and the time axis is in picoseconds (10^{-12} s). The data is taken from [2].

where n_2^e(water) and n_2^e(solute) are the instantaneous electronic contributions of the water and solute, respectively, to the measured OKE signal. In the second term on the rhs of (9.5), $n_2^o(i)$ refers to the ith component of the slow or molecular (non-electronic) response to the measured OKE signal, which is decaying with a time constant $\tau(i)$. Thus, there are purely electronic responses and molecular responses for both solvent and solute; in fact there are two nuclear or molecular contributions from the solute, $n_2^o(ii)$ and $n_2^o(iii)$, with time constants $\tau(ii)$ and $\tau(iii)$, respectively. Thus the mass of data represented by these indirect measurements will be fitted to an equation (9.5), which has five unknowns, and which will be determined simultaneously by the fitting process. To achieve anything from such an exercise, we require, at least five, equations that relate the values of interest to the experiment. As it happens, because we are able to vary the time between the pumping and probing pulse, we can generate the required expressions, and the various unknowns may be found by fitting the data to the model function; all the fitting being done by least-squares methodology. Figures 9.2 and 9.3 show the best fits results for the various solutions investigated.

In figures 9.2 and 9.3, we see the measured relaxation of water molecules and solvated ions after their polarizability has been ordered by the pulse or pump laser. As it happens, the dynamics of water molecules in liquid water is one of the biggest areas of research in chemical physics (it is a field of endeavour that has applications in modern medicine and in global warming), and there are many theories as to how the water molecules interact in the condensed phase (see [3]). However, the signal to noise of the data represented in figure 9.2(a) did not allow us to differentiate between these various mechanisms for the relaxation of water molecules in liquid water, Consequently, we left our measured relaxation constant as $\tau(i) = 700 \pm 50$ fs, and made no comment as to why the property has this value; it has an error of 50 in 700 or 7%, which is to be expected as the effect being measured is weak, even though many thousands of measurements contributed to the final result.

In the case of sodium chloride solution and hydrochloric acid, it was found that the data could be well represented by invoking only one parameter to describe the molecular relaxation, with a relaxation time constant to represent the contribution of the solute ions to the OKE; component (ii), $\tau(ii) \sim 1.5$ ps. However, with the nitrates ($NaNO_3$ and HNO_3), the nitrate anion is planar and geometrically anisotropic and, consequently, the observed OKE was larger than for the chlorides, and an additional molecular contribution (iii) also characterized by an exponential decay was required to fully represent the data. The time constant, $\tau(iii)$, was found to be a function of the sample concentration.

It is the much larger signal arising from the polarization of the nitrate anions in the nitrate systems that gave us the opportunity to fit the data to three parameters as opposed the data derived from the chloride systems; in the chlorides there was a smaller OKE and the data was, consequently, noisier and so fitting with one parameter or with three parameters made no difference to the residuals (we will encounter the problem of trying to over-fit data in the next chapter).

It is the quality of the data, the signal to noise of the experiment that allows one to fit multi-variate models to the data from a single experiment. As it happened, in the nitrates there was the instantaneous electronic contribution (clearly seen in the data) and two molecular relaxation components, (*ii*) with time constant $\tau(ii)$ of about 2.2 ps and (*iii*) with relaxation constant $\tau(iii)$. The first of these two contributions was found to be largely independent of solute concentration, while the second term was found to be strongly dependent on solute concentration, and was interpreted as the onset of a glassy-like state in the solution where there is not enough water to shield the bare ions from each other.

It is not my intention to give a detailed explanation of the type of information that may be derived from OKE experiments on aqueous ionic solutions. But merely to point out that such experiments are possible and they are typical of an enormous class of indirect measurements (for example, most particle physics experiments, all neutron scattering experiments and all crystal structure determinations involve indirect measurements), where what you are interested in comes from fitting a specific model to a large body of measured data. But even in these experiments, if the signal generated is not great, as in the OKE of pure water and in solutions of HCl and NaCl, then the signal to noise may not be sufficient to allow one to discriminate between different models, each with several unknowns, for what may be happening in the fluid, or to isolate the contributions of the various components present in the solution to the overall measured signal. In the latter case, one would need to undertake additional experiments using additional samples and varying other parameters; for example, the sample temperature.

There is always a trade-off between how many parameters you put in the theoretical model you have constructed of what it is that you think is going on in your experiment, and the quality of the measured data. If the signal to noise is poor (figure 9.2(*a*)) do not try and fit the data with a complex model; the solutions you derive from the fitting procedure (and your computer will generate solutions!) will be unstable, and will themselves be subject to large uncertainties. As the signal to noise improves (as for example in figure 9.2(*b*)) you may attempt to quantify ever more subtler effects in the data (that may or may not be there).

References

[1] Buckingham A D 1956 *Proc. Phys. Soc.* **B69** 344
[2] Santa I, Foggi P, Righini R and Williams J H 1994 *J. Phys. Chem.* **98** 7692
[3] Williams J H 2015 *Order from Force* (San Rafael, CA: Morgan and Claypool)

Chapter 10

Data fitting and elephants

With four parameters I can fit an elephant and with five I can make him wriggle his trunk.

Attributed to John von Neumann by Enrico Fermi

10.1 Introduction

As a young research professor in 1953, the eminent British-born theoretical physicist and mathematician, Freeman Dyson (born 1923) made a trip to Chicago to visit his mentor, Enrico Fermi. Dyson and his group had been calculating meson–proton scattering cross-sections using a theory (that is, a model) of the strong nuclear force known as pseudoscalar meson theory, which had been published by Fermi. Dyson was impressed and excited by the apparent agreement between his theoretical computations and the available experimental data measured by Fermi in Chicago. This agreement had emboldened Dyson to visit Fermi to present his results. Sadly for Dyson, Fermi on viewing the results was unimpressed. Dyson naturally asked why Fermi was not impressed by the apparent agreement between theory and experiment. Fermi responded by asking a question, 'To reach your calculated results, you had to introduce arbitrary cut-off procedures that are not based on solid physics or on solid mathematics ... How many arbitrary parameters did you use for your calculations?' Dyson's response was four, to which Fermi famously commented, 'I remember my friend Johnny von Neumann (1903–57) used to say that with four parameters I can fit an elephant and with five I can make him wriggle his trunk'. The interview between Dyson and Fermi was over [1].

By saying this, Enrico Fermi was not insulting his enthusiastic young guest, but he was pointing out a fundamental truth of data analysis. That is, with enough arbitrary parameters, it is possible to find a curve that goes through just about any set of data points. This statement is perhaps not true for a set of truly random numbers, but apparently it is true for a set of data that when graphed would

doi:10.1088/978-1-6817-4433-9ch10

resemble the outline of an elephant, with a big fat body, distinct legs, a tail and a trunk. Then, of course, if you find an equation containing several adjustable parameters that links all the data points in your complex, exotic data set, you are naturally led to the assumption that you now understand the physics behind the force (s) that is(are) generating your odd looking set of data. Of course, there is no natural phenomenon that could generate such an outlandish dataset, the outline of an elephant, but Fermi, and presumably, von Neumann, were using this weird and wonderful idea to instruct their listeners in the dangers of putting all your trust in data fitting algorithms. When interpreting data; particularly, the results of theoretical calculations made to interpret experimental results (particularly, experimental results with appreciable uncertainties), one should never lose sight of the underlying physics.

10.2 Regression analysis

Regression analysis is the most widely used statistical tool for the investigation of relationships between variables. Usually, an investigator is seeking to learn if there is a causal effect of one variable upon another: for example, what is the relationship between income and education, or is there a temperature-independent term that contributes to the measurement of the quadrupole moment via electric-field gradient induced birefringence?

To explore such questions, the investigator assembles data on the underlying variables of interest and employs regression to estimate the quantitative effect of the causal variables upon the variable that they (may or may not) influence. The investigator also assesses the statistical significance of the estimated relationships; that is, the degree of confidence that the true relationship is close to the estimated relationship.

The techniques of regression analysis were first used by the 19th Century British biologist and statistician, and cousin to Charles Darwin, Francis Galton (1822–1911). Galton was a pioneer in the application of statistical methods to measurements in many areas of biology; specifically, in analyzing data on relative sizes of parents and their offspring in studies of plants and animals. He famously observed the following: a larger-than-average parent tends to produce a larger-than-average child, but the child is likely to be less large than the parent in terms of its relative position within its own generation. If the parent's size is x standard deviations from the mean within its own generation, then one could predict that the child's size will be $r \cdot x$ standard deviations from the mean within the set of children of those parents, where r is a number less than 1 (the correlation between the size of the parent and the size of the child). Such observations may be found for physical measurements throughout biology, and in humans for most measurements of cognitive and physical ability; and it led to the horrors of eugenics in the 20th Century. For Galton, regression had only this biological meaning, but his work was later extended to a more general statistical context. For the statisticians, the joint distribution of the response and explanatory variables is assumed to be Gaussian.

Francis Galton termed this phenomenon a 'regression towards mediocrity', which today is expressed as a 'regression to the mean' and it is an inescapable fact of life.

Your children can be 'expected' to be less exceptional (for better or worse) than you are. Your score on a final exam in a course can be 'expected' to be less good (or bad) than your score in the mid-term exam, relative to the rest of the class. The key word here is 'expected'. This does not mean it's certain that regression to the mean will occur, but that's the way things tend to go in terms of the probability.

In data analysis, errors and residuals are two closely related and often confused measures of the deviation of an observed value of an element of a data set from its 'theoretical value'. The error of an observed value is the deviation of the observed value from the (unobservable) true value of a quantity of interest (for example, a population mean), and the residual of an observed value is the difference between the observed value and the estimated value of the quantity of interest (for example, a sample mean). The distinction is particularly important in regression analysis.

Suppose there is a series of observations from a univariate distribution and we wish to estimate the mean of that distribution. In this case, the errors are the deviations of the observations from the population mean, while the residuals are the deviations of the observations from the sample mean.

A statistical error is the amount by which an observation differs from its expected value, the latter being based on the whole population from which the sample was chosen randomly. For example, if the mean weight of a group of individuals is 80 kg, and a randomly chosen individual has a weight of 85 kg, then the 'error' is 5 kg; however, if the randomly chosen individual has a weight of 78 kg then the 'error' is now −2 kg. The problem is that the expected value, being the mean of the entire population, is typically unobservable, and hence the statistical error cannot really be observed.

A residual (or fitting deviation), on the other hand, is an observable estimate of the unobservable statistical error. Consider the previous example of weights and suppose we have a random sample of n people. The sample mean could serve as a good estimator of the population mean. Then we have the difference between the weights of each person in the sample and the unobservable population mean is a statistical error, whereas the difference between the weight of each person in the sample and the observable sample mean is a residual.

The sum of the residuals within a random sample is necessarily zero, and thus the residuals are necessarily not independent. The statistical errors on the other hand are independent, and their sum within the random sample is almost certainly not zero. In regression analysis, the distinction between errors and residuals is subtle and important, and leads to the concept of studentized residuals. Given an unobservable function that relates the independent variable to the dependent variable; for example, a straight line, the deviations of the dependent variable observations from this function are the unobservable errors. If one runs a regression on some data, then the deviations of the dependent variable observations from the fitted function are the residuals.

In the statistical modelling of experimental or observational data, regression analysis is the means of estimating the relationships among the variables involved in the measurements. It has evolved into a standard procedure that is today often encountered as a 'black box' technology often already programmed into the computers that record experimental data. The technique includes many different ways of modelling and analyzing several variables, when the focus is on the relationship between a dependent variable and one or more independent variables (or predictors). More specifically, regression analysis helps one understand how the typical value of the dependent variable (or criterion variable) changes when any one of the independent variables is varied, while the other independent variables are held fixed. Most commonly, regression analysis estimates the conditional expectation of the dependent variable given the independent variables; that is, the average value of the dependent variable when the independent variables are fixed. In all cases, the estimation target is a function of the independent variables called the regression function. In regression analysis, it is also of interest to characterize the variation of the dependent variable around the regression function which can be described by a probability distribution.

Let us consider an example: the influence of education upon income, for which we take a large sample of individuals of similar age and ask them how long they spent in full-time education and what is their present income. Clearly, when graphed this data will generate a plot with an enormous range of uncertainty or scatter. But the question that has to be answered (and is often asked): is there a relationship between education level and income? The graph will suggest that higher values of education tend to result in higher incomes, but the relationship is not perfect; and it would be clear that knowledge of education level does not suffice for an accurate prediction of income. To apply regression analysis to this particular problem requires that we first hypothesize that earnings for each individual are determined by education, and by a collection of other factors (race, locality, sex, profession, etc.) that we term contributing noise.

Then, we write a hypothesized model relationship or equation between education (E) and income (I) as; for example,

$$I = \alpha + \beta E + \varepsilon,$$

where α is a constant amount (what one earns with zero education), β is the coefficient relating how an additional year of education influences income (assumed to be positive), and ε is the noise term representing other factors that influence earnings (factors that are unobservable, or at least unobserved). The variable I is the dependent or endogenous variable and E is termed the independent, explanatory, or exogenous variable. The parameters α and β are not observed, and regression analysis is used to produce an estimate of their value, based upon the information contained in the data set.

What we have hypothesised is that there is a straight line or linear relationship between E and I. Thus somewhere in the cloud of data points on our graph we expect to find a line defined by the equation $I = \alpha + \beta E$. The task of estimating α and β is equivalent to the task of estimating where this line is located on the axes? The answer depends in part upon what we think about the nature of the noise term ε. If we have reason to believe that ε is zero, then the line to be fitted to the scattered data points will lie somewhere in the cloud of data points. If ε has a large value (positive or negative), then the fitted line will lie above or below the cloud of points.

In figure 4.1, we see a set of measured data points that were plotted in a manner so as to answer the question; is there a temperature-independent contribution to the measured quadrupole moment as determined by electric-field gradient induced birefringence? That is, what is the best line through these data points, and is there a finite value of the intercept on the x-axis?

Regression analysis is used to search for the best line through a set of data points, thereby providing values for the coefficients in a model equation. This is the line that reflects the estimated error for each data point as the vertical distance between the value of, for example, I along the estimated line $I = \alpha + \beta E$ (generated by putting the actual values of E into our hypothesised equation or model) and the true value of I for the same observation; that is the difference between theoretical values of the variable and actual measurements of the variable.

With each possible line that might be superimposed upon the data set, a different set of estimated errors will result. Regression analysis then chooses among all the possible fitted lines by selecting the one for which the sum of the squares of the estimated errors is a minimum. This is termed the minimum sum of squared errors (minimum SSE) criterion. The intercept of the line chosen by this criterion provides the value of α, and its slope provides the value of β.

Why should we choose our line using the minimum SSE criterion? We can readily imagine other criteria that could be used; for example, minimizing the sum of errors in absolute value. One virtue of the SSE criterion is that it is straightforward to use. When one expresses the sum of squared errors to find the values of α and β that minimize it, one obtains expressions for α and β that are easy to evaluate using only the observed values of E and I in the data set. Continuing with our example, imagine that we have data on E and I for a number of individuals indexed by j. The actual value of I for the jth individual is I_j, and its estimated value for any line with intercept α and slope β will be $\alpha + \beta E_j$. The estimated error is thus $I_j - \alpha - \beta E_j$. The sum of squared errors is then $\sum_j (I_j - \alpha - \beta E_j)^2$. Minimizing this sum with respect to α requires that its derivative with respect to α be set to zero, or $-2\sum_j(I_j - \alpha - \beta E_j) = 0$. Minimizing with respect to β requires $-2\sum_j E_j(I_j - \alpha - \beta E_j) = 0$. We now have two equations with two unknowns that can be solved for α and β. But computational convenience is not the only virtue of the minimum SSE criterion; it also has some attractive statistical properties under plausible assumptions about the noise term, ε.

Many techniques for carrying out regression analysis have been developed. Familiar methods such as linear regression and ordinary least-squares regression are parametric, in that the regression function is defined in terms of a finite number of unknown parameters that are estimated from the available data. Nonparametric regression refers to techniques that allow the regression function to lie in a specified set of functions, which may be multi-dimensional. The earliest form of regression was the method of least-squares, which was published by Legendre in 1805 and by Gauss in 1809, and which was famously used to re-interpret the data obtained by Delambre and Méchain during the metric survey of the 1790s (see chapter 2).

The actual form of the regression analysis to be used depends on the form of the data generating process, and how it relates to the regression approach being used. Since the true form of the data-generating process is generally not known (it is well characterized in a direct experiment, but is less well-defined in an indirect experiment), regression analysis often depends to some extent on making assumptions about this process. These assumptions are sometimes testable if a sufficient quantity of data is available.

Regression models generally involve the following variables:

- The unknown parameters, denoted as β, which may represent a scalar or a vector.
- The independent variables, $\mathbf{X}$.
- The dependent variable, Y.

In various fields of application, different terminologies are used in place of dependent and independent variables. A regression model relates Y to a function of $\mathbf{X}$ and β.

$$Y \approx f(\mathbf{X}, \beta), \text{ cf. } I = \alpha + \beta E + \varepsilon.$$

The approximation is usually formalized as $E(Y \mid \mathbf{X}) = f(\mathbf{X}, \beta)$. For regression analysis, the form of the function f must be specified. Sometimes the form of this function is based on information about the relationship between Y and $\mathbf{X}$ that does not rely on the data. If no such information is available, a convenient form for f is chosen.

Assume that the vector of unknown parameters β is of length k. In order to perform a regression analysis one must provide information about the dependent variable Y:

1. If N data points of the form $(Y, \mathbf{X})$ are observed, where $N < k$, most classical approaches to regression analysis cannot be performed: since the system of equations defining the regression model is underdetermined; there is not enough data to recover β.
2. If $N = k$ data points are observed, and the function f is linear, the equations $Y = f(\mathbf{X}, \beta)$ can be solved exactly rather than approximately. This situation reduces to solving a set of N equations with N unknowns (the elements of β), which has a unique solution as long as the $\mathbf{X}$ are linearly independent. If f is nonlinear, a solution may not exist, or many solutions may exist.
3. The most common situation is where $N > k$ data points are observed. In this case, there is enough information in the data to estimate a unique value for β that best fits the data in some sense, and the regression model when applied to the data can be viewed as an overdetermined system in β. This is the situation with the optical Kerr effect discussed in the previous chapter.

In case 3, the regression analysis provides the tools for:

- Finding a solution for unknown parameters β that will, for example, minimize the distance between the measured and predicted values of the dependent variable Y (this is the method of least-squares).
- Under certain statistical assumptions, the regression analysis uses the surplus of information to provide statistical information about the unknown parameters β and predicted values of the dependent variable Y.

Consider a regression model which has three unknown parameters, β_0, β_1, and β_2. Suppose an experimenter performs several measurements all at the same value of independent variable vector $\mathbf{X}$ (which contains the independent variables X_1, X_2, and X_3). In this case, regression analysis fails to give a unique set of estimated values for the three unknown parameters; the experimenter did not provide enough information. The best one can do is to estimate the average value and the standard deviation of the dependent variable Y. Similarly, measuring at two different values of $\mathbf{X}$ would give enough data for a regression with two unknowns, but not for three or more unknowns.

If the experimenter had performed measurements at three different values of the independent variable vector $\mathbf{X}$, then regression analysis would provide a unique set of estimates for the three unknown parameters in β. In the case of general linear regression, this statement is equivalent to the requirement that the matrix $\mathbf{X}^{\mathrm{T}}\mathbf{X}$ be invertible.

When the number of measurements, N, is larger than the number of unknown parameters, k, and the measurement errors ε_i are normally distributed, then the excess of information contained in $(N - k)$ measurements is used to make statistical predictions about the unknown parameters. This excess of information is referred to as the degrees of freedom of the regression.

Classical assumptions for regression analysis include:

- The sample is representative of the population for inference prediction.
- The error is a random variable with a mean of zero, conditional on the explanatory variables.
- The independent variables are measured with no error. If this is not so, modelling may be done instead using errors-in-variables model techniques.
- The independent variables (predictors) are linearly independent; that is, it is not possible to express any predictor as a linear combination of the others.
- The errors are uncorrelated; that is, the variance–covariance matrix of the errors is diagonal and each non-zero element is the variance of the error (see the sections on propagation of uncertainty in chapters 5 and 6).
- The variance of the error is constant across observations. If not, weighted least-squares or other methods might instead be used.

These are sufficient conditions for the least-squares estimator to be useful; in particular, these assumptions imply that the parameter estimates will be unbiased, consistent, and efficient in the class of linear unbiased estimators. It is important to note, however, that actual data rarely satisfies the assumptions; that is, the method may be used even though the assumptions are not valid.

In linear regression, the model specification is that the dependent variable, y_i is a linear combination of the parameters (but need not be linear in the independent variables). For example, in simple linear regression for modelling n data points there is one independent variable: x_i, and two parameters, β_0 and β_1, we define a straight line (as we did above for education and income):

$$y_i = \beta_0 + \beta_1 x_i + \varepsilon_i \quad \text{for} \quad i = 1, \ldots, n.$$

In multiple linear regression, there are several independent variables or functions of independent variables. Adding a term in x_i^2 to the preceding regression generates a parabola,

$$y_i = \beta_0 + \beta_1 x_i + \beta_2 x_i^2 + \varepsilon_i \quad \text{for} \quad i = 1, \dots, n.$$

This is still termed linear regression, although the expression is quadratic in the independent variable x_i. In both cases, ε_i is an error term and the subscript i indexes a particular observation. For the straight line case; given a random sample from the population, we can write the population parameters and obtain the sample linear regression model:

$$\underline{y}_i = \underline{\beta}_0 + \underline{\beta}_1 x_i.$$

The residual, $e_i = y_i - \underline{y}_i$, is the difference between the value of the dependent variable predicted by the model, $\underline{y}_i$, and the true value of the dependent variable, y_i. One method of estimation is ordinary least-squares, and as we saw above this provides estimates that minimize the sum of squared residuals, SSE, or:

$$SSE = \sum_i e_i^2.$$

Minimization of this function generates a set of simultaneous linear equations in the parameters of interest, which are solved to yield the parameter estimators, $\underline{\beta}_0$, $\underline{\beta}_1$.

In the case of simple regression, the formulas for the least-squares estimates are:

$$\underline{\beta}_1 = \sum(x_i - x_{\text{mean}})(y_i - y_{\text{mean}}) \Big/ \sum(x_i - x_{\text{mean}})^2 \quad \text{and} \quad \underline{\beta}_0 = y_{\text{mean}} - \underline{\beta}_1 x_{\text{mean}}$$

where x_{mean} is the mean (average) of the x values and y_{mean} is the mean of the y values.

Under the assumption that the population error term has a constant variance, the estimate of that variance is given by:

$$\underline{\sigma}_\varepsilon^2 = \frac{SSE}{n - 2}.$$

This is called the mean square error (MSE) of the regression. The denominator is the sample size reduced by the number of model parameters estimated from the same data, $(n-p)$ for p regressors or $(n-p-1)$ if an intercept is used. In this case, $p = 1$ so the denominator is $n-2$.

The standard errors of the parameter estimates are given by

$$\underline{\sigma}_{\beta 0} = \sigma_\varepsilon \sqrt{(1/n)} + \left\{ x_{\text{mean}}^2 \Big/ \sum(x_i - x_{\text{mean}})^2 \right\}$$

$$\underline{\sigma}_{\beta 1} = \sigma_\varepsilon \sqrt{1 \Big/ \sum(x_i - x_{\text{mean}})^2}.$$

Under the further assumption that the population error term is normally distributed, one can use these estimated standard errors to create confidence intervals and conduct hypothesis tests about the population parameters.

Once a regression model has been constructed, it is necessary to confirm the 'goodness of fit' of the model, and the statistical significance of the estimated parameters. Commonly used checks of goodness of fit include the R-squared, analyses of the pattern of residuals and hypothesis testing. Statistical significance can be checked by an F-test of the overall fit, followed by t-tests of individual parameters.

Interpretations of these diagnostic tests are based on the model being assumed. Although examination of the residuals can also be used to invalidate a model, the results of a t-test or F-test are sometimes more difficult to interpret if the model's assumptions are violated. For example, even if the error term does not follow a normal distribution, in small samples the estimated parameters will not follow normal distributions and complicate inference. With relatively large samples, however, the central limit theorem can be invoked such that hypothesis testing may proceed using asymptotic approximations.

Regression models usually predict a value of the Y variable, given known values of the $\mathbf{X}$ variables. Prediction within the range of values in the data set used for model-fitting is known as interpolation. Prediction outside this range of the data is termed extrapolation. Performing extrapolation relies strongly on the assumptions upon which the regression analysis is based. The further the extrapolation goes outside the data, the more room there is for the model to fail due to differences between the assumptions and the sample data, or the true values. When performing extrapolation, one should accompany the estimated value of the dependent variable with a prediction interval that represents the uncertainty. Such intervals tend to increase in size rapidly as the values of the independent variable(s) move further outside the range covered by the observed data.

However, this does not cover the full set of modelling errors that may be made: in particular, the assumption of a particular form for the relation between Y and $\mathbf{X}$. A properly conducted regression analysis will include an assessment of how well the assumed form is matched by the observed data, but it can only do so within the range of values of the independent variables available. This means that any extrapolation is reliant on the assumptions being made about the structural form of the regression relationship. It is best to use all available knowledge in constructing a regression model.

But whatever the caveats, regression analysis is widely used for prediction and forecasting; not only for estimating the values of variables outside of the area where related measurements have been made, but there is a substantial overlap with the design of predictive texting (there are also many economists who seek to apply such techniques to the data shown in figure 4.5 with a view to making short-term extrapolations). However this can lead to false relationships, so caution is advisable; for example, correlation does not imply causation (for example, which came first, the economic recession or increasing levels of unemployment).

10.3 Over-fitting data

As we have seen, if we determine n distinct values of x, and the corresponding values of y, it may be possible to find a curve that goes precisely through all n points (x,y) when graphed. This can be done by setting up a system of equations and solving them simultaneously. But this is not what regression methods are designed to do.

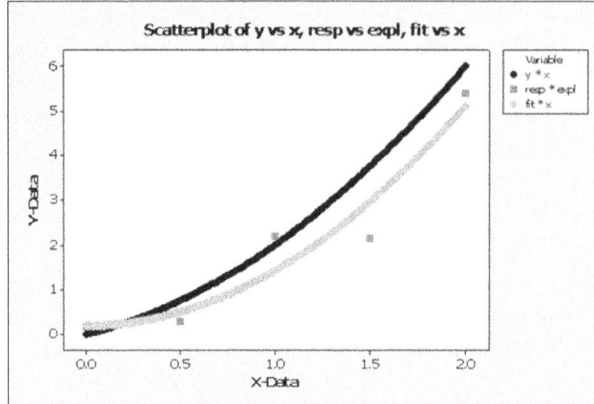

Scatterplot of y vs x, resp vs expl, fit vs x

Most regression methods (for example, least-squares) estimate conditional means of the response variable given the explanatory variables. They are not expected necessarily to go through all the data points. Indeed, the residuals between the curve calculated from your model and the measured data is the means of expressing the uncertainty of the measurements, and there is always uncertainty. For example, with one explanatory variable X and response variable Y, if we fix a value x of X, we have a conditional distribution of Y given $X = x$; for example, the conditional distribution of the incomes of individuals who have had five years of education. This conditional distribution has an expected value (the population mean), which we will denote $E(Y|X = x)$; that is, the mean-income of people with five years of schooling, x. This is the conditional mean of Y given $X = x$. It depends on x, as $E(Y| X = x)$ is a function of x.

In least-squares regression, one of the model assumptions is that the conditional mean function has a specified form. Then we use the data to find a function of x which approximates the function $E(Y|X = x)$. This is different from finding a curve that goes through all the data points. Consider the following example to illustrate this point.[1] We will use simulated data: five points sampled from a joint distribution where the conditional mean $E(Y|X = x)$ is known to be x^2, and where each conditional distribution $Y|(X = x)$ is normal with standard deviation 1. Least-squares regression is used to estimate the conditional means by a quadratic curve $y = a + bx + cx^2$; that is, least-squares regression, with $E(Y|X = x) = \alpha + \beta x + \gamma x^2$ as one of the model assumptions, to obtain estimates a, b, and c of α, β, and γ, respectively, based on the data.

The above graphic shows:
- the five data points in red (one at the left is mostly hidden by the green curve)
- curve $y = x^2$ of conditional means (black)
- graph of the calculated regression equation (in green).

---

[1] Taken with permission and thanks from the teaching material of Professor Martha K Smith, University of Texas; ma.utexas.edu/users/mks/statmistakes/.

Note that the points sampled from the distribution do not lie on the curve expressing the means (black). Notice again, that the green curve is not exactly the same as the black curve, but is close to it. In this example, the sampled points are mostly below the curve representing the means. Since the regression curve (green) was calculated using just the five sampled points (red), the red points are more evenly distributed above and below the green curve than they are in relation to the curve representing the means (black).

Remember that in the real world, we would not know the conditional mean function (black curve); and in most problems, would not even know in advance whether it is linear, quadratic, or something else. Thus, part of the problem of finding an appropriate regression curve is trying to discover what kind of function it should be. Continuing with the example, if we (perhaps naively) try to obtain a 'good fit' to the data by trying a quartic (fourth degree) regression curve; that is, using a model assumption of the form $E(Y|X = x) = \alpha + \beta_1 x + \beta_2 x^2 + \beta_3 x^3 + \beta_4 x^4$, we obtain the following graphic.

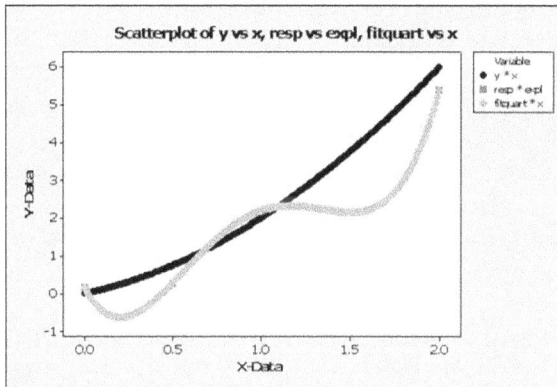

Here we barely see any of the red points, because they are all on the calculated regression curve (green). We have found a regression curve that fits all the data, but it is not a good regression curve, because what we are really trying to estimate by regression is the black curve (curve of conditional means). And for that, we have done a poor job; we have made the mistake of overfitting. We are well on the way to fitting an elephant, so to say. If we had instead tried to fit a cubic (third degree) regression curve (that is, using a model assumption of the form $E(Y|X = x) = \alpha + \beta_1 x + \beta_2 x^2 + \beta_3 x^3$), we would get something more wiggly than the quadratic fit and less wiggly than the quartic fit. However, it would still be overfitting, since (by construction) the correct model assumption for these data would be a quadratic function.

10.4 Avoiding over-fitting

As with most problems in statistics and data analysis, there are no hard and fast rules that guarantee success; it all depends upon the problem under investigation and the quality and quantity of the data. However, there are some guidelines, which apply to many types of statistical models. Of premier importance is the quality and quantity of your data. If you are gathering data (especially through an experiment), be sure to optimise the design and set-up of the apparatus before beginning to take data. There are no generally agreed methods for relating the number of observations needed to be made in an experiment to the number of independent variables in your model for that experiment, which will allow you to decide *a priori* whether your model is any good or will lead to useful results. This is also the problem with the design of clinical trials, as we saw in chapter 7. In their textbook, P I Good and J W Hardin [2] offer the following conjecture or rule of thumb on an appropriate sample size: 'If m points are required to determine a univariate regression line with sufficient precision, then it will take at least m^n observations and perhaps $n!m^n$ observations to appropriately characterize and evaluate a regression model with n variables'. That dataset could become very large, very quickly. Happily for the OKE measurements described in the previous chapter, it is relatively straightforward to acquire large quantities of data by varying the experimental conditions; for example, the delay time between the orienting and probing laser pulses.

Having mentioned von Neumann's elephant, the question naturally arises as to whether or not this is at all possible. That is, is it possible to model and define the silhouette of an elephant with four parameters, and then make the simulation wriggle its trunk by the addition of a fifth parameter?

Earlier, we came across a crestfallen Freeman Dyson who had had a few of the facts of a physicist's life pointed out to him by Enrico Fermi. I must confess to having heard much the same thing when working as a physicist in the Institut Laue-Langevin in Grenoble, the largest European facility for research involving neutrons and neutron scattering. It has to be said, however, that neutron scattering, particularly quasi-elastic neutron scattering, is a field where it is only possible to make progress by fitting the data measured by scattering neutrons from samples containing hydrogen to a range of complex (and sometimes opaque) models. One morning, while discussing with a visiting scientist the scattering data that had accumulated overnight, the question arose as to how many parameters should we use in attempting to fit the data to the model we thought appropriate. No sooner had we started talking, than someone in the group quoted the comment about the elephant. Apparently, these critics of data fitting always quote von Neumann's aphorism with a smile, even though they themselves have probably never done any data fitting more sophisticated than eye-balling a straight line through a cloud of points. But the story of the elephant will not go away, which probably means that there is some truth in it. So how easily can you fit the shape of an elephant?

James Wei was the first to demonstrate that something like the shape of an elephant could be generated by computer fitting. In 1975 [3], Wei took the outline of an elephant defined by 36 points (see figure 10.1) and used least-squares Fourier series fits of the form $x(t) = \alpha_0 + \sum_i \alpha_i \sin(it\pi/36)$ and $y(t) = \beta_0 + \sum_i \beta_i \sin(it\pi/36)$ for $i = 1,\dots N$ to try and represent the defined outline of the elephant. Figure 10.1 shows the fits Wei obtained for $N = 5, 10, 20$ and 30; he stopped at a thirty parameter fit, which has a certain similarity to the 36-point image of an elephant which Wei had

Figure 10.1. Fitting the outline of an elephant. These images are taken from the fitting of James Wei in 1975 [3]. One can clearly see the evolution of the outline of the elephant; from an egg to the embryo and then to the elephant, but Wei was using a lot more than four parameters in his fit—and there is no wiggling trunk. (*a*) James Wei's starting point, an elephant defined by 34 points (not a simulation, but a drawing). (*b*) A least-squares fit with five terms—more of an egg than an elephant. (*c*) A least-squares fit with 10 terms. (*d*) A least-squares fit with 20 terms—an embryo of an elephant? (*e*) A least-squares fit with 30 terms—now something like an elephant.

drawn to be his ideal, but as we can see, it is not too good, and is using significantly more than the five parameters mentioned by von Neumann.[2]

This work by Wei is humorous and insightful. The standard image against which Wei was seeking to generate an image was itself pretty crude. Perhaps the target image should have been a digitized photograph, but again this is only an approximation to an elephant, and we can clearly see that the generated fits do not really merit any further investigation. This straightforward exercise should, however, encourage thinking about the nature of reality, about what constitutes a true model and the 'right answer', and about the merits of 'approximate models'. Perhaps there is no such thing as the 'perfect fit', in the same way that there is no such thing as the 'right answer' or a measurement free of uncertainty.

A more recent attempt to parameterize an elephant also makes use of Fourier series. The outline of the beast is best described by a set of points $(x(t), y(t))$, where t is a parameter that can be described as the time elapsed while moving along the path of the animal's outline [4]. If motion along the animal's contour is uniform, t defines the length of an arc or a curve of the animal's body. Expanding $x(t)$ and $y(t)$ in Fourier series gives

$$x(t) = \sum_k A_k^x \cos(kt) + B_k^x \sin(kt)$$
$$y(t) = \sum_k A_k^y \cos(kt) + B_k^y \sin(kt)$$

where A_k^x, B_k^x, A_k^y and B_k^y are the coefficients of the expansion; k defining the term in the expansion and x and y labelling the expansion of the x and y function, respectively. Using such Fourier expansions it is possible to analyze shapes by

---

[2] Interestingly, the least-square fits of James Wei given in figure 10.1 suggest the gestation of the elephant, from an egg to the animal. Perhaps this is a new take on the now largely discredited hypothesis that in developing from embryo to adult, animals go through stages resembling successive stages in the evolution of their remote ancestors; that is, the Theory of Recapitulation ('*ontogeny recapitulates phylogeny*') of the German biologist Ernst Haeckl.

tracing the boundary and calculating the coefficients in the expansion. By truncating the expansion, the derived shape is smoothed. The question is how many truncated parameters are required to give the derived curve the outline of an elephant? From Fourier analysis, we know that $k = 0$ corresponds to the centre-of-mass of the perimeter, $k = 1$ corresponds to the best fit ellipse. The higher components trace out elliptical corrections analogous to Ptolemy's epicycles (see en.wikipedia.org/wiki/Deferent_and_epicycle). Mayer *et al* were able to generate the image seen in figure 10.2, using the expansion coefficients given in their paper.

Here the authors have only used the required four parameters, but the final result is a comical cartoon of an elephant. The real part of the fifth parameter is the 'wriggle coefficient' which determines the x-value where the trunk is attached to the body. The imaginary part of this parameter is used to make the cartoon more 'life-like' by giving it an eye.

In another approach, the programmers formed a second elephant by reflection of the distal portion of the proboscis of the first elephant, parallel to the x-axis. The wiggling was then implemented by linear interpolation between the primary and secondary elephants, using a sinusoidally varying mixing coefficient. The virtual

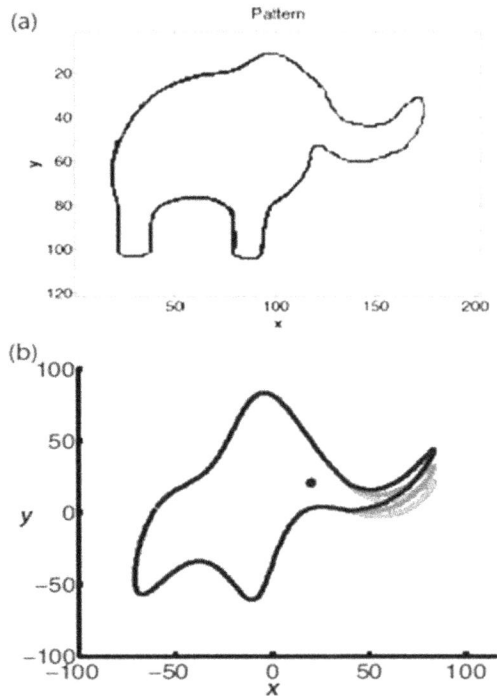

Figure 10.2. These images are taken from [4]. In these images, the real part of the fifth parameter is the 'wriggle coefficient' which determines the x-value where the trunk is attached to the body. The imaginary part of this parameter is used to make the cartoon more 'life-like' by giving it an eye. All the images in this figure and figure 10.1 are cartoon-like, but are they elephant-like?

elephant may be observed here: http://theoval.cmp.uea.ac.uk/~gcc/projects/elephant/ And interactive graphics at http://demonstrations.wolfram.com/FittingAnElephant/ and http://kourentzes.com/forecasting/2016/02/08/how-to-fit-an-elephant/ allows one to observe directly the data fitting to this strange and instructive shape.

Success in the analysis of real data, and the resulting inferences often depend crucially on the choice of a best model. Data analysis in those fields where researchers have a liking for multi-variate model fitting (I am sure that there are economists who are quite happy to construct hugely complex models to try and fit the type of data displayed in figure 4.5 as the potential rewards are too great to ignore) should be based on a strict application of Occam's razor, *Essentia non sunt multiplicanda praetor necessitatem* (entities are not to be multiplied beyond necessity); that is, represented by an accurate approximation of the data in hand. This fitting should not be thought of as a search for the 'true model'. Modelling and model selection can all too easily become an art, rather than an aid to science. And when fitting large data sets, one needs to be aware that it is possible to fall into a minimum in the fitted surface (as defined by the value of your residuals), perhaps even a deep minimum, but which is not the true minimum.

Enrico Fermi gave Freeman Dyson a tutorial on the dangers of over-fitting; something which all of us who have undertaken complex indirect experiments have been guilty of at one time or another. It is a cautionary tale in physics, warning us not to lose sight of the physics (and the concept of measurement uncertainty) when programming and fitting data. As it happens, Dyson took Fermi's point and changed his area of research. As Dyson commented over half a century later, '*I am eternally grateful to him, for destroying our illusions and telling us the bitter truth*' [1]. But now we do know that it is possible, with only four complex parameters to generate the vague outline of an elephant (with a contemporary aesthetic), and with a fifth complex parameter one can control both the position of the trunk and the position of an eye.

Further reading

There are some very useful videos on the Wikipedia entry for curve fitting, that demonstrate in real time what a scientist experiences when he or she is trying to fit his measured data to some existing model that they have programmed: https://en.wikipedia.org/wiki/Curve_fitting).

There are many standard university texts on the do's and don'ts of data fitting, and how to avoid the elephant traps of over-fitting data; for example [2, 5] (these texts, or parts of them are also available as free-access pdf files). The online encyclopaedia, Wikipedia is also a good source for further information (see entries on statistical fitting, regression analysis, curve fitting, goodness of fit, and errors and residuals). I can also recommend the following site at the University of Texas: www.ma.utexas.edu/users/mks/statmistakes/Types.html

References

[1] Dyson F 2005 *Nature* **427** 297

[2] Good P I and Hardin J W 2006 *Common Errors in Statistics (And how to avoid Them)* (New York: Wiley)

[3] Wei J 1975 *Chemtech.* February 128–9

[4] Mayer J, Khairy K and Howard J 2010 *Am. J. Phys.* **78** 648–649

[5] Ryan T 2009 *Modern Regression Methods* (New York: Wiley)

www.ingramcontent.com/pod-product-compliance
Lightning Source LLC
Chambersburg PA
CBHW081544220326

41598CB00036B/6564